Monkey Business

Monkey Business

A History of
Nonhuman Primate Rights

Erika Fleury

Library of Congress Control Number: 2013910787

ISBN: 1490384022
ISBN-13: 978-1490384023
07162013v1

This book is dedicated to Grandma Ruth,
who would have been proud.

Contents

Preface

This book is designed to document progress that has been made in the struggle for nonhuman primate rights. As I researched the subject, I learned that many early efforts to support a major cause—or a fight for rights or protections for a group—have been met not only with a defensive wall to hurdle but, even worse, with outright public apathy. In this case, whenever I explained the concept of my book, I would hear over and over, "Primate *what?* Primate *rights?*" After making a comparison with the animal rights movement (a different but related and much more well-known concept), I would then often be asked what primates were. I knew I had important work ahead of me.

It is my hope that documenting the history of primate rights will make the cause even more real and important. Earlier works have covered some of this material; however, recent changes to laws—and, thankfully, to public opinion about the proper treatment of nonhuman primates—encouraged me to write and collect this information into one cohesive history. I wanted to assemble it as quickly as possible, for I knew that the sooner I could get it out in the world—bound, printed, and ready to educate readers—the more quickly our culture might progress toward legal freedom for our closest animal relatives, the nonhuman primates.

Erika Fleury

Thank you to:

My family, for supporting my dreams.

Those who directly helped to make this book a reality, including (but not limited to):
April Truitt and the Primate Rescue Center, for giving me a chance and continuing to help me so many times in my travels throughout the world of primatology; my brother-in-law, Kyle Aaron, for his amazing cover design talents; and my night-owl of an editor, Deanna Brady, for turning my book into something I would want to read.

My daughter, the beautiful Leila Jaye Fleury, for wisely delaying her birth and allowing this book to be completed.

And mostly, thank you to my husband, Teague Fleury, for always inspiring me in our shared mission to protect little things.

Prologue

"Blood is thicker than water," promises an old proverb, and the actions of most of us prove this to be true. People naturally place a high level of importance on the safety, happiness, and comfort of their relatives. It is generally understood that individuals are more likely to value the overall quality of life of their kin above the well being of absolute strangers.

Even in the animal world, members of various species seem to know that certain other members are part of their extended family and, as such, are more deserving of patience, favors, and protection from threats to their welfare. This is how bloodlines are perpetuated. Living creatures have a Darwinian instinct to act in such a way as to best preserve themselves and their descendants. Without this drive, their offspring would not live long enough to reproduce, and the family line would die out.

When this preservation instinct is acted upon, a species usually continues to evolve and grow ever more prosperous in its environment. As higher numbers of relatives survive and reproduce, families expand at an exponential rate. This means that, at some point, a thriving bloodline could grow large enough that it would be difficult for its members to recognize some of their own more distant relations.

How thin can blood become before it appears to be no more than water? Where do we draw the line between family and the rest of the world? Is there a point at which we stop feeling connected to distant relatives? Certainly our mothers, grandfathers, aunts, and nephews are our relatives. What about their relatives by marriage or remarriage? What about second and third cousins? How do we consider their children?

With each dilution of the genetic bond to us, individuals will likely be considered less and less as family members. At a genetic distance they will typically garner less of our sympathy and concern and seem less deserving of our protection.

Most human societies place a high value on human life. Cultures may assign levels of inherent worth to specific human lives, depending on such characteristics as race, gender, religion, ethnicity, behavior, and sexual orientation, but individuals and groups typically align and feel united with those most like themselves. As a corollary to this tendency, humans tend to value human life over that of other species. Past the dividing lines of family, friends, acquaintances, and strangers is a boundary where the human species abuts others, and humans generally consider that line to be sacrosanct.

Human beings are members of a vast and varied order of intelligent and highly specialized animals known as primates. Considering that there are more than 230 other distinct species of primates, however, it becomes clear that human beings are but a blip on the primate radar.

Despite that minority status, we often refer to our closest relatives in this order as nonhuman primates, which is a way to define them as "not like us" but which can be misleading in terms of the group as a whole. To refer to all other primates as nonhuman, simply in order to separate them from the less than one percent of the primate population that is human, is akin to referring to almost all apples as "non Macintosh," when only a small fraction of the apple world is actually of the Macintosh variety. In fact, this is only one example, among many, of the way humans categorize the world's other species in purely anthropocentric terms.

The best way to illustrate the biological relationships between humans and other species is through the universally accepted system of ranking the living world developed by Carl Linnaeus in the mid 1700s. The Linnaean Classification System establishes distinct points at which various groups of living beings branched off evolutionarily from their common ancestors, with each branching producing an increasingly distinctive group.

Each of these branches is delineated with a new category so that the path a given species followed to evolve its particular traits can be traced. For example, the vast class *Mammalia* includes all mammals, and within this class resides the order

Primates, to which belong the more than two hundred species of apes, monkeys, and prosimians. Moving down the primate chain a bit farther, through suborder *Anthropoidea,* infraorder *Catarrhini,* superfamily *Hominoidea,* and all the way through family *Hominidae* (great apes) and subfamily *Homininae,* we reach the point at which the genus *Homo* (human beings) finally separates from the other primates.

Within the primate family, the genetics of humans are most similar to the bonobo (genus *Pan* and species *paniscus),* even more so than to the chimpanzee (genus *Pan* and species *troglodytes).* The genus name refers to the Greek god Pan, a herder who had goat legs but the torso of a man. The name may have been chosen for chimpanzees and bonobos in order to reflect how close they really are to humans. Seeming to have one foot in the animal kingdom and one in the human, chimpanzees and bonobos appear to be the missing link between man and the rest of the natural world.[1]

Before the designation of the bonobo as a distinct species, it had previously been believed that the chimpanzee was humanity's closest relative. (Estimates and conclusions varied, placing the degree of similarity between human and chimpanzee DNA between 95 and 99 percent.) It was later decided that, due to a more recent split from our common ancestor, human beings are slightly more closely related to bonobos than they are to chimpanzees. Variations in the percentage of relatedness exist because of the methods each researcher employs and the degree to which each may consider samples similar or different.

Even with such a small window of genetic dissonance, a 2006 University of California study showed that most of those regions of genetic difference are not responsible for coding genes but appear to be located near the centers that tell the DNA when to code genes and how to do so. Those types of non-coding genes evolve more rapidly than coding genes because they do not have to answer to selective pressure, which explains why most of the differences between humans and chimpanzees lie in these faster-to-evolve bits of DNA. As Katherine Pollard, one of the authors of the 2006 study, explained in a Science Daily article, "The differences between chimps and humans are not in our proteins, but in how we use them."[2]

One estimate claims that, of the 100,000 genes responsible for substantive parts of human anatomy, we differ from other primates by only a few hundred genes[3]; yet, those few hundred genes result in major differences that make humankind quite distinct from other primates. It's not the bricks of the house but the architecture that comprises the majority of the differences between humans and our closest primate relatives. Our foundation is the same.

Viewing the Linnaean system of classification of the animal world as a whole, it quickly becomes apparent that human beings are truly a small fraction of the *Primates* order—a fraction that became distinct only relatively recently. By some estimates, primates have been around for 85 to 95 million years, but it was only approximately six to eight million years ago that human beings evolved away from the common ancestry we share with the other great apes, and more specifically with chimpanzees and bonobos.

The mental capabilities of human beings have evolved faster than those of other species, which allowed humans to create tools to manipulate their environments, protect them from predators and the elements, and eventually control other species despite extreme disparities in their physical strength and prowess. We humans are able to do this simply because of the way our brains evolved. This stroke of good fortune makes humans' dominance over the natural world seem rather arbitrary and, quite frankly, based on the luck of the evolutionary draw.

Despite the possibly accidental evolution of humanity's dominance over other species, people often take their power for granted. The most common view in Western culture, for example, is of humans as both physically and morally separate from the natural world, including animals. As a result, many human societies have historically found ways to manipulate their environments—and the animals that dwell there—so as to maximize their wealth potential. As a consequence, other primate species are used by humans in ways too numerous to count, including some that are not necessary for human survival.

Examples of such abuse seem especially cruel when these species' intelligence is taken into consideration. For instance, infant primates in the wild are often snatched from their mothers after witnessing them being shot by poachers. These traumatized babies are then shipped across the globe and sold as pets to people who typically do not know how to care for them properly. Other primates are bred in concrete laboratories, destined to lead joyless lives as they are injected with the latest drugs on trial, sometimes never even seeing another member of their own species. Still others are dressed in costumes, taught to mimic human behaviors such as dancing or riding a bicycle, and then paraded on television or under the big top for the amusement of humans.

Despite the fact that these primates' brains have not developed as highly as humans' have, they are quite intelligent. Certain primate species are capable of long-term memory (pleasant or unpleasant), sharing language and culture (whether their own or humans'), and relying on vast, complex social networks (sadly missing from the solitary lives many primates lead in captivity). Unfortunately their mental ability and physical dexterity are exactly what makes them so desirable to humans for purposes of exploitation.

Nonhuman primates are enough like humans to be mesmerizing but are also different enough that people can distance themselves from their pain. This seems to have been the case virtually forever, but public opinion seems to be changing over time. The more people learn about other primates, the more they respect them, and the more they wish nonhuman primates to be protected from further human involvement.

Science is beginning to find alternatives to primate research. Laws are changing regarding keeping primates as pets or entertainers, laws that benefit the health and safety of humans and captive primates alike. Sanctuaries are now being supported financially and encouraged as safe havens where primates who have been removed from the wild can live out their lives in as peaceful and natural a manner as possible, away from becoming sources of revenue for people any longer. Humans are finally beginning, at least to some extent, to evolve away from the kind of humor that encourages laughing at an ape in a costume, and perhaps soon people will no longer find it funny when an intelligent, sensitive creature is debased and defiled.

1. The History of Animal Rights

But let this be your invariable rule, everywhere, and at all times, to do unto others as, in their condition, you would be done unto. [4]
- Revered Humphrey Primatt

In order to comprehend fully the state of affairs that comprises the movement for nonhuman primate rights, it is important to understand its foundations in the history of rights for all animals. This is a long and complex tale that is still evolving, but in the legends of the past we see portents of the future, and there is hope that this evolution will reflect the continuous uplifting of the state of nonhuman primates.

Of those working to improve the conditions of animals in relationship with humans, there are two major schools of thought. The animal rights movement supports interspecies equality and generally maintains the belief that animals should not be used by humans for food, clothing, entertainment, research, or indeed any other purpose. Furthermore it holds that animals' basic rights should remain intact and unaffected by the lives of humans, because their very existence on Earth implies an inherent worth. By this logic, any member of the animal kingdom is entitled to certain rights of life and liberty and protection from harm, as well as the freedom to live as naturally as possible.

The animal welfare movement is related to the animal rights movement but is a bit watered down, as it were, allowing for some usage of animals by humans, albeit in a humane manner and measured on a sliding scale of ethical consideration. Some animal welfarists would support animal rights in a perfect world but believe that animal welfare is a more practical starting point from which to make real change in modern culture, perhaps in the hope that momentum will eventually lead to more complete animal rights.

Animal welfarists seek to eliminate gratuitous animal suffering, although animal rightists would say that all the ways animals are currently used by humans are technically gratuitous. In their view, using animals in bioresearch or as entertainment or even in food production is unnecessary; humankind can survive (some would say with an even better quality of life) by using exclusively non-animal products.

One important issue sometimes raised about animal welfarism is its potential for inefficiency. Creating more humane living conditions for animals is important, but by creating more humane conditions within a system that considers animals a means to an end, welfarism indirectly supports practices that do not have animals' best interests at heart. In fact, although there are currently more animal welfare laws on the books than ever before, more animals are actually being used for human gain than ever before.[5]

The debate regarding animals' rights versus their welfare, however, was a later development in the history of what we now place under the umbrella of animal rights. Specifically in the United States, the beliefs behind what we know as animal rights were first recorded as an unassuming mention in a colonial document.

Since the beginning of civilization, animals have typically been considered items of property that belong to humans. In many European languages, for instance, the words for property and capital derive from such words as chattel and cattle, since using animals as items of exchange was one of the first ways to conduct business in the absence of a uniform currency.[6]

When the United States was in its infancy, European people settling the country brought with them these pre-established views of animals and how they related to human life. The first known legal precedent regarding animals came relatively early in American history. In 1641, The General Court of Massachusetts developed the Body of Liberties, which was essentially a list of the do's and don'ts of conducting a proper Puritan life. Amongst the one hundred statutes was Libertie 92, which stated, "No man shall exercise any Tirranny or Crueltie towards any bruite Creature which are usuallie kept for man's use."[7] Of note are the last four words of Libertie 92, "kept for man's use," which illustrates the view commonly held of the time: that domesticated animals were the property of humans.

It's not surprising, then, that any early laws made regarding the treatment of animals were clearly created in order to protect proprietary objects (as opposed to considering animals as beings with their own inherent worth, an idea that wouldn't be put forth until much later). Prior to this Colonial time period, pre-Christian and Christian thought asserted man's dominion over God's kingdom, and although a few philosophers and scholars such as Saint Thomas Aquinas and Saint Francis of Assisi took it upon themselves to ponder and discuss how animals should be treated, the overwhelming majority of people and cultures viewed animals as objects put on the earth to aid humanity's existence. Renaissance thought held up man as an example of supreme perfection, and in general a person who might speak out in concern of animal suffering was both rare and subject to intense scrutiny and skepticism.

During the development of what would become The United States of America, Puritan ideals valued tolerance, morality, and "standards (implying) ascertained immutable truths that were right and good in themselves and that must be observed to the letter."[8] The abuse of animals would by definition have been antithetical to a peaceful, ethical Puritan life. Although Libertie 92 was more a way of protecting the rightful property of individuals than a declaration of the inherent value of nonhuman life, it did ultimately gain some protections for animals living in captivity and is marked as the first legal precedent on behalf of animals living in the United States.

The British Reverend Humphrey Primatt wrote his *A Dissertation on the Duty of Mercy and Sin of Cruelty to Brute Animals* in 1776 and asked that his fellow citizens "see that no brute of any kind...whether entrusted to thy care, or coming in thy way, suffer thy neglect or abuse. Let no views of profit, no compliance with custom, and no fear of ridicule of the world, ever tempt thee to the least act of cruelty or injustice to any creature whatsoever. But let this be your invariable rule, everywhere, and at all times, to do unto others as, in their condition, you would be done unto."[9] These lofty and pure ideas were not legitimized through association with any other philosophies of the day, although another Englishman would soon create a name for himself by addressing one of the more often debated topics in the philosophical world.

In 1789, Jeremy Bentham's *An Introduction to the Principles of Morals and Legislation* was the first published work to examine many of the themes important to modern animal rights, such as equal considerations of interests across species, as well as the ethics of animal suffering. His writing gained much attention and was heavily relied upon by the movers and shakers of the animal rights community. This Englishman directly opposed the principle proposed by Rene Descartes in 1637 that assigned moral consideration based on proof of intelligence.

Cartesian philosophy was grounded in the premise that man was rational, man had a soul, and no other beings could sufficiently be proven to exhibit rational thinking; therefore, Descartes deduced, only man had a soul and deserved moral concern. The physical movements of animals led Descartes to compare them to machinery, devoid of feeling or mental processes, and to conclude that their needs, desires, or inherent worth need not be considered by the ethical man. Strangely enough, Cartesian philosophers did not question the absence of logic involved when a man simply assumes that only humans have mental processes, when it is a human who is defining the very principle. This inherent conflict of interest naturally creates major problems for animals, who cannot converse in a mutual language with the Cartesian philosopher about his faulty reasoning.

In contrast, Jeremy Bentham countered, "...a full grown horse or dog is beyond comparison a more rational, as well as a more conversable animal, than an infant of a day, or a week or even a month, old. But suppose the case were otherwise, what would it avail? The question is not, Can they *reason*? nor Can they talk? but, Can they suffer?"[10] Bentham was able to overlook the arbitrary issue of linguistic capability of nonhumans, and in doing so he questioned the heart of the matter, which was equal consideration of interests. He relied on the fact that most people, including Americans of the late 1700s, would agree that an injured animal also feels pain, as illustrated by the physical and vocal reactions that all mammals share in response to pain. Therefore, his followers reasoned, the typical reactions of animals to pain meant that causing pain was bad and should be avoided whenever possible. Bentham's words and sentiment proved to be timeless and inspirational to future animal rights and welfare activists, and his quote is still widely referenced in discussions of animal rights today.

Increasingly supported and inspired by a similar social movement in Europe, Americans who sympathized with the feelings of animals worked to ensure that the progress earned across the Atlantic would one day make the pilgrimage to the growing United States. The nineteenth century would prove to be an important one for animal welfare, as it was then that much of the legal groundwork was discussed and put into action, and there was a transformation in the way people viewed animals. Increasingly, they were no longer considered unfeeling objects and were instead considered sentient beings, replete with their own rights to comfort.

A bill called Martin's Act was passed in England in 1822, which made it illegal to abuse a wide variety of equine and bovine animals. Martin's Act found it "expedient to prevent the cruel and improper Treatment of Horses, Mares, Geldings, Mules, Asses, Cows, Heifers, Steers, Oxen, Sheep, and other Cattle" and made it illegal to "wantonly and cruelly beat, abuse, or ill-treat"[11] those creatures. Unfortunately the act was rarely enforced, and in 1824 the Society for the Prevention of Cruelty to Animals was established in England partly in order to further define and strengthen Martin's Act.

Back in North America, it took a few more years for similar rulings to be established. The first anti-cruelty law was passed in Maine in 1821, and New York and Massachusetts established their own anti-cruelty laws in 1828 and 1835,

respectively[12]. Other states would soon follow. Although most of
the laws continued to be based on the opinion that animals were
nothing but human property to be protected against loss, each
new regulation paved the way for future laws, and each increased
the moral considerations of individual animals in their own right.

A breakthrough came in 1859 with the publishing of
Charles Darwin's revolutionary research and book *The Origin of
Species*. Evidence of evolution within the animal kingdom linked
lofty human beings with the snuffling swine, the braying donkey,
and even the scuttling crab climbing across the ocean floor. A
physical connection among species proved that, rather than
masterful stewards of the universe, humans were merely one of
many. Humans and other animals now had common relatives,
and it suddenly seemed somewhat presumptuous to assume that
humans were preordained to rule the animal kingdom.
Evolutionary connectedness implied a comparable allocation of
physical traits, with emotional traits not too far behind. Those
fighting for protection of animals now had the reliable backbone
of science to defend them from the increasingly sensationalist
media and a public that was reluctant to change.

It didn't take long for Darwin's theories to be applied—or
misapplied, depending on one's perspective—to various contexts.
If the human being represented a mere segment of the animal
kingdom that had evolved more highly by natural selection,
mused the cynics, then human dominance over other animals was
completely natural and logical and should be embraced.
Additionally they figured that since humans had developed
superior brains and reasoning power, they should assert these
traits in order to follow their obviously predestined path of
supremacy.

The animal rights movement countered this view with the
reminder that human society functioned on many deeper levels
than simply that of the strong governing the weak. Activists
referred to democracy, education, justice, philanthropy, and
altruism. They asserted that picking and choosing certain
scientific theories to rationalize unfair treatment of other species

was ignorant and overly simplistic and that this notion of justified domination was not what they believed Darwin had intended his studies to prove.

Another argument gleaned from Darwinism put forth the rationale that violence between species was exhibited naturally in the animal world. This ultimate application of the survival of the fittest theory led people to ask why, then, should humans not physically harm other species, when they are perfectly capable of doing so and probably designed to do so, as well? The retort from activists was that, in civilized societies, human survival did not depend on the killing of other animals for food or shelter or even on using them for transportation. According to historian Diane Beers, when humans physically harmed animals, "...motivations other than survival, such as power, love, revenge, and greed, frequently distinguished it. The human capacity for savagery was unique in its potential for unjustifiable evil, and for humanitarians, few acts illustrated this truth better than the reprehensible, unprovoked abuse of animals."[13]

An interesting twist was put on the discussion by animal rightist Rosa Abbot, who agreed that humans were indeed subject to Darwinist evolution and drives, but this would include not merely physical but also spiritual evolution. Abbot's humans were meant to grow past their earthly urges and act instead from a higher moral ground.[14]

As topics relating to humane treatment of animals became more visible in the public eye, people were naturally motivated to group together to organize their efforts and produce results for the cause. The first work of organizing supporters of animal rights in the US was done in Philadelphia in 1860 and was principally accountable to the persistence, social connections, and fundraising ability of two ladies of society: Caroline Earle White and S. Morris Waln[15]. When the Civil War began in 1861, however, that conflict took precedence in American minds for the next few years.

Because the Northern states' denunciation of the slave trade could be correlated with the humane ownership of animals, after the war ended, "for those interested in animal protection, the victory inaugurated hopeful discussions among some reformers that perhaps society would consider an even broader definition and application of natural rights that would include other species as well."[16] The denial of rights based on race seemed eerily similar to the denial of rights based on species. Abolitionists and suffragists alike sympathized with the nascent animal rights movement and found many similarities in its fight for equality. The parallels were clear: Those in power took advantage of and asserted their dominance over the voiceless, and a suppressed group of individuals with distinct interests, goals, and personalities was denied essential freedoms.

As the United States grew increasingly industrialized after the Civil War, more and more people were drawn to work in the factories of burgeoning cities. The future of commerce lay in mass production, and the gears and cogs that spit out the goods to feed ever-growing populations required human bodies for maintenance and management.

For the first time in history, industrialization brought about a way for large numbers of animals to be processed into products swiftly and in a seemingly nonstop stream. Innovative killing techniques, assembly methods, and refrigeration and preservation mechanisms expanded the amount of food and by-products that could be extracted from each animal carcass. People were no longer living on farms, communing with nature and experiencing life with animals, but were instead beginning to view animals as simply another replaceable piece in a great machine. In the stockyards of Chicago, as Upton Sinclair described it in *The Jungle*, "There were groups of cattle...the stream of animals was continuous; it was quite uncanny to watch them, pressing on to their fate, all unsuspicious—a very river of death. Our friends were not poetical, and the sight suggested to them no metaphors of human destiny; they thought only of the wonderful efficacy of it all."[17]

Not every American was impressed with the new technology. An appreciable number of citizens decried the destruction of nature and lamented the increasing speed of production of the mechanized modern world. Thoreau wrote,

"The twelve labors of Hercules were trifling in comparison with those which my neighbors have undertaken; for they were only twelve, and had an end; but I could never see that these men slew or captured any monster or finished any labor.[18]

In what may have at times seemed a daunting environment, activists continued to fight on behalf of animals, demanding more humane treatment and systems for their care. After seeing animals coldly abused and harmed while traveling in Europe, Henry Bergh founded the American Society for the Prevention of Cruelty to Animals (ASPCA) in April of 1866.[19]

During one particular episode, Bergh had been able to convince a carriage driver to stop whipping his exhausted donkey. He realized that persistence and direction could have a profound effect on the systems negatively affecting animal treatment in the United States. After collecting much support from his wealthy, high-ranking friends back home in New York City to fund his ASPCA, Bergh declared it a person's moral and religious duty to treat animals with respect and dignity.

Groups similar to the ASPCA began to appear in various cities around the country. Slowly, the animal rights movement was being taken more seriously and gaining a larger following. Part of the reason was the persistent reliance on education of the public and bringing to light questions about supposed human superiority and control over the natural world. The wealthy upper class tended to have more leisure time to ponder such ethical issues, and that same group could also afford to feed the movement dollars and help it to reach a growing American populace.

Progress continued to be made in Europe, and advances there on behalf of animal welfare must have seemed like rays of hope to America's animal protectionists. Although Americans had worked hard to separate themselves from English rule, there were still many cultural and legislative precedents and connections with England, and many legal rulings established there would eventually make an appearance across the ocean. England's Cruelty to Animals Act of 1876 was established to address the increasingly debated topic of vivisection (experimentation on live animals). This act required that all scientists practicing vivisection be licensed, and it prohibited numerous experiments on the same un-anaesthetized subject. The act also granted the government permission to draw the boundaries of what is deemed "acceptable" research.[20] The vivisectionist community organized such a fight against the bill that it was generally thought of as rather unsuccessful. As it turned out, the issues and complications inherent in the English fight for animal rights were but a foreshadowing of similar battles ahead in America's vivisectionist community.

The ASPCA was responsible for a landmark law passed in April 1866 (the same month the organization was established), which "stipulated that the state had to classify and prosecute any act of cruelty toward any animal as a misdemeanor offense, regardless of issues of ownership."[21] The ASPCA was now responsible for enforcing the new law in New York City, and, like other groups that started locally and later became nationalized in order to more efficiently influence federal lawmakers, it quickly became a national organization so as to best enforce the law in other locations, as well.

Despite the changes in animal treatment slowly being introduced into American society, it remained difficult to gain major ground within the political system, and early activists quickly grew frustrated with laws that touched only the surface of the rights to which they believed animals were entitled. Large populations of animal protectionists banded together, in groups such as the ASPCA in New York, Pennsylvania, and Massachusetts, and even spread west to Illinois, Minnesota, and California.

The first group of animal activists made good use of the analogy between animal servitude and that of human slaves, because increased sympathy could be gleaned from that movement. Anna Sewell's novel *Black Beauty* was the first book told from the perspective of a victimized animal, and it made all too vivid the abuse experienced by service animals.

There was less interest in animal rights in the southern part of the United States, for many reasons. It was not helpful there to refer to the causes of the suffragists or abolitionists, for instance, simply because these were not popular movements in the post–Civil War South. Southern societies were more agrarian and rural than those in the North, and animals were still relied upon for the well-being and financial support of many families and communities. Some groups for the protection of animals were formed in Southern states, but they lagged in popularity and were not truly active until later in American history.

In areas where the societies were popular, a wide variety of reforms were embraced. Education remained a valued tool to instruct adults and also reach children who would soon grow to comprise society. It was hoped that conversion of youngsters to the cause would result in a more compassionate society within just a few generations. Animal rights groups embraced the establishment of shelters, especially in urban areas, where such sanctuaries (if they existed at all) were typically overfilled and under-funded. Volunteers helped to staff and maintain the shelters, as well as to support ordinances that dealt with the abundance of homeless animals in urban areas. Additional projects included building water fountains for workhorses, opening the first retirement homes for work animals, and creating and expanding public veterinary systems.

The early organizations in support of animal rights were plagued with many of the issues that any developing organization might be prone to: lack of focus and lack of public understanding (or blatant misunderstanding) of the cause, as well as specialization or division among group members. The early defenders of animal rights found many points on which to agree or disagree: which animals deserved protection, from what, and how best to create systems of protection. In addition, the sudden consideration of the welfare of animals would threaten many major systems in modern culture, from the railroads to the federal government and even the diet of the average omnivore.[22]

Those prosecuted under the new anti-cruelty legislation often thought themselves victims of a silly joke, and unfortunately the legal system frequently failed to support the cause in defense of animals. The public was slow to embrace

Henry Bergh's ASPCA as a policing body, but in 1866, its first year of enforcement, the group prosecuted 119 cases in New York City alone, with 66 convictions.[23] Slowly the laws involved became more defined and resulted in the establishment of direct-enforcement bodies, and the protection of animals grew to be an increasingly accepted part of the legal system in New York City and spread out to the rest of the United States.

Change is often difficult. In retaliation against the enforcement of new business practices, and also as a defense mechanism to protect entrenched values and systems of the industrialized world, propaganda and public opinion derided the animal rights movement and painted the new activists as anti-progress, anti-science, anti–public health, and even anti–organized government. In making the average person believe that animal rights could not conceivably coexist with the interests of the human race, polarization was generated, and the average person was asked to choose between the "human" vs. "animal" consideration of interests.

When faced with such a black-and-white question, many people chose "human"; however, the ASPCA found that with every news article mocking their beliefs or ridiculing their latest arrests, their membership soared. In an early application of the motto "Any publicity is good publicity," the groups protecting animals continued their actions with optimism and a firm conviction in their beliefs, regardless of the public ridicule they might momentarily face.

Despite promoting greater awareness of the cause of animal rights, not all the attention garnered was helpful. Some vociferous defenders of the status quo labeled the animal activists as insane or imbalanced. Recognizing that wealthy and educated women were the first supporters to swell the ranks of the ASPCA and its local counterparts, opponents of the cause questioned the femininity and morality of group members, claiming that they nursed animals and abandoned their human families. In an apparent attack ploy created and utilized by some who felt threatened by defenders of animal rights, the diagnosis of a psychological condition dubbed zoophilpsychosis was established in 1909. Women were supposedly more susceptible to this so-called disease, symptoms of which included a disproportionate concern and worry for animals[24]. Understandably, zoophilpsychosis failed to become a recognized medical condition in the long term.

A public relations method used by early animal protection agencies was to show how protecting animals would ultimately be good for human beings, as well. It was noted that "the early campaign to enact more-humane slaughter methods, for example, attached cruelty concerns to decidedly anthropocentric concerns about tainted meat and human health."[25] Abuse of animals was also discussed as a precursor to violence perpetrated against humans.

The reasoning behind this approach was simple: At any point in world history, the problems deserving of charity are numerous. Even in times of the greatest plenty there is always poverty, and with it come the associated charities that vie for attention and funding from those who can afford to contribute. The argument can be posed that as long as there is human suffering in the world, that should take precedence over the needs of any animal; thus, the animal rights movement linked their concerns with what was best for humans and, as a result, ultimately gained more support for their cause.

As animal rights groups like the ASPCA gained more power and grew increasingly vocal with a public that took them ever more seriously, the businesses accused of cruelty to animals often stopped their practices, not because they saw the error of their ways but because they feared the negative public opinion and resulting potential loss of business. Women made up the majority of the groups in support of animal rights, and women were also the managers of their households at the time. This meant that when women decided, for instance, to boycott fur products on behalf of dwindling seal populations, their actions had the potential to affect the fur industry greatly, and tremendous attention could be attracted to their cause.

The first animals to benefit from the new anti-cruelty systems were those most visible in the culture of the time, such as workhorses and mules. Laws were passed to protect the welfare of the carting animals, mentioning weight limits to loads of goods and/or human passengers towed on carts, the regular care and maintenance of the animals, and humane treatments for disabled, elderly, or injured animals. An ambulance system was created to transport animals to receive urgent medical attention—something that wasn't yet available for human beings[26]! Although workhorses would soon be a thing of the past because of the invention and subsequent popularity of the automobile, the early protections and legal victories in regard to workhorses would pave the way for protections related to many of Earth's creatures, including those raised as pets and for food, clothing and decoration, entertainment and sports, and research.

Established in 1877, the American Humane Association grew specifically out of concern for the plight of stockyard animals[27]. Humane changes were needed in meat processing plants because the processing system was built for speed and quantity of product, not safety and quality. Overworked employees, little-to-no sanitary rules or enforcement, and overcrowded, often diseased cattle made for a grisly scene. Animals were being torn apart while still conscious; diseased, filthy meat was ground up and sold as food; and uneducated employees were frequently injured as they handled dangerous weapons and tools. All this happened around a prime source of the nation's food supply, and animal rights groups recognized that not only was this a travesty for the animals involved, it was also affecting human well-being.

For the first time, animal defenders chose to depend on diplomacy rather than aggravation, hoping that by working with stockyard leaders they could avoid confrontation and come to agree on mutually beneficial terms that would assist both humans and animals[28]. Discussions focused on improving designs and methods used when transporting animals, as well as how animals in the meatpacking plants were handled, killed, and stored. The decision to include the meatpacking industry drew ire from more radical animal rights organizations and limited the focus and the actions that the American Humane Society was able to take. In return, however, the meatpacking industry helped fund the organization, which was then able to advertise on a national level.

Although issues pertaining specifically to domestic animals were not of primary importance to animal rights groups, consideration of the large numbers of homeless dogs and cats living in America's cities soon spread. Many people believed that the systems in place to deal with (or dispose of) homeless animals left much to be desired. For example, New York City regularly drowned cages-full of homeless animals, simply because they had no established way to care for them[29].

Philadelphia animal welfare societies created the first animal shelters, and similar animal sanctuaries soon opened in other areas of the country. Although many shelters had no choice but to end the lives of their animal residents simply to reclaim limited space, they focused on humane methods of killing that would cause the least amount of suffering to animals who had already suffered much in their lives. The first no-kill shelter, Bide-A-Wee, was eventually established in 1903 in New York, replicating shelters already in existence in Europe[30].

Animal protection groups also worked to end blood sports, a term that described various potentially lethal practices, from cockfighting to bull baiting, that were immensely popular early in the last century. Through sheer determination and in spite of the blatant violence and threats of physical harm that confronted officers, rings were broken up and improvements were made. These sports eventually waned in popularity, in part due to positive support of animal defenders in the press. Nonetheless, many blood sports popular with the upper crust of American society, such as pigeon shoots and foxhunts, continued to be held in high esteem[31].

Conservationists argued that the sports helped control excessive populations of the target animals and maintained a proper equilibrium in nature. Hunting, they believed, also showed a respect for nature and was simply a by-product of man's placement at the top of the natural food chain. As Diane Beers notes, "Throughout the controversy, middle- and upper-class participants in the sport waged a relatively effective public relations drive in which they contrasted themselves as conscientious stewards of nature with what they perceived as overly sentimental animal lovers."[32] In response, Ernest Thompson Seton published *Lives of the Hunted* in 1901, which sympathized with hunted animals and did much to promote the agendas of animal protection agencies.

The usage of animals for entertainment purposes, which in the late 1800s and early 1900s was mostly limited to zoos and circuses, was also highly scrutinized. Although zoos were an affordable way for people living lives of the new industrialism to commune with nature and see animals from distant lands, the real problem for animal defenders arose with animal training programs. The public was never exposed to the cruel and inhumane tactics used by early animal trainers and handlers. Audiences were blinded by the lights and sequins, never seeing the pain and suffering going on behind the big top. The ASPCA was the only organization with enough manpower and gumption to try to take on the well-funded and media-savvy P.T. Barnum, but despite its history of successes, it was outwitted and ended up getting burned in terms of public opinion[33]. The group was never able to prove publicly that inhumane conditions existed in circuses.

Animal skins used in fashion were immensely popular in the early 1900s, especially furs. Advances in trapping allowed hunters to decimate various populations, driving some nearly to extinction. The first laws did not end trapping practices but merely required hunters to monitor their traps and set limits on the amount of time an injured animal could languish in a trap before a hunter was required to remove it[34].

Also drawing attention were the negative effects on bird populations from the demand for feathers to decorate the highly popular and intricately fashioned hats of the day. The method of tearing wings off the bodies of adult birds not only caused them to starve to death slowly but resulted in their young, dependant offspring perishing as well. Animal protection groups added the cause of imperiled birds to their list of concerns, and groups such as the Audubon Society were established to study and protect bird species. The Lacey Act of 1900 was passed to restrict importation and protect native bird populations in the US, and additional legislation soon followed. Birds were slowly becoming respected and admired, and bird watching became a popular pastime. Again, activists revealed the cruelty in former practices and began to change the nation's behavior for the better.[35]

The great victory established with ending the millinery feather trade proved something pivotal: that education could indeed bring about change. When made aware of the pervasiveness of needless animal suffering and when provided with alternatives that could avoid causing more unnecessary animal suffering, people would often choose the path of least pain. Animal protection groups began distributing brochures, booklets, and leaflets and sponsoring lectures and essay contests, all in order to educate the public and particularly the children, who would one day grow up to become the next generation of law-making adults.

In 1868 the first humane-treatment magazine, *Our Dumb Animals,* was published by George Angell and inspired the creation of many similar publications. In 1889 Angell founded the American Humane Education Society to promote and monitor the cause of humane-treatment education[36]. Clubs were formed to further involve America's children, starting with The Juvenile Society for the Protection of Animals, founded by Caroline Earle White. George Angell later created a similar youth group, The Bands of Mercy, modeled after a British organization, which flourished at the national level, with many local societies forming their own branches. Such youth groups were wildly successful in America, crossing geographical and even racial lines. In 1915, the third week of May was deemed Be Kind to Animals Week, with various local events and awards ceremonies planned in celebration. The event continued to be celebrated annually, growing in popularity and scale.[37]

The early1900s were a period of relative inactivity for animal rights groups, not only because it would have been difficult to maintain the rapid growth and expansion of the previous century but because World War I distracted many people with more pressing concerns. The failure of some groups to reach their goals (such as ending vivisection) painted all humane-treatment groups with the same broad brush, distracting from the very real progress that had been achieved by other, more moderate organizations. The Progressive movement had begun, drawing attention to more human-centered problems. Progressivism did incorporate certain themes of animal protection, but in more generalized terms, such as labeling animal abuse as bad, if only because abuse of living things in general was deemed bad.

Changes in society and the new ways animals were being used meant that groups had to realign their focus—away from the dwindling practice of horse-drawn carts, for example, and towards animals used in the film industry. However, the plight of homeless dogs and cats remained as pervasive as ever, and many groups found themselves overwhelmed with that one issue alone, which led to a loss of funds and attention for other issues related to animal protection. Despite this, activists continued to educate and reach people by whatever methods they thought would work best, from radio chats in the East to revivalist-style tent meetings in the South.[38] Persistence and creativity was their strength and backbone.

Although it was a distraction for the movement, World War I would also prove to create novel ways to draw attention to the plight of animals. It was recognized that the horses and dogs used in battle were frequently subject to inadequate veterinary care when injured, and they were lucky to get any medical attention at all. To fill this void, the American Red Star, an animal emergency relief service, was approved for use by the War Department in 1916, and local divisions were soon successfully raising money to support the cause.[39] Public opinion of this patriotic faction of animal rights was highly favorable. (World War II changed things a bit, however, as machines began to replace animals on the battlefield. Due to this shift, American Red Star would later revise its mission to focus more on the rescue and recovery of animals used in defense programs and those affected by natural disasters.[40])

Around the same time, popular nature writer Jack London grew disgusted with the abuse involved in trapping and training wild animals for circus shows and wrote *Jerry of the Islands* and *Michael, Brother of Jerry,* books that provided firsthand accounts of what a life in circus training was like for a wild animal living out of his element and forced to perform at the risk of personal injury. The public reacted very strongly to his words, and after his death in 1916, Jack London clubs were formed and gained thousands of members. There was such an outcry that the Ringling–Barnum and Bailey circus even temporarily stopped using live animals in their shows.[41] The arrival of the Great Depression, however, meant that the circuses had to resort to using any means possible to draw crowds and earn profits, so the use of animals as circus performers continued.

When Americans fell in love with the silver screen early in the century, it quickly became imperative to monitor the use of animals in the film industry. Serious public outcry finally arose after a scene staged for the 1939 film *Jesse James* required a horse to be driven off a cliff and plummet to its death in a pool of water. The American Humane Association was then assigned the task of approving animal treatment on all film lots, where efforts proved more successful than with circus animals.[42]

The romantic view of nature initially popularized by Jack London was further propelled by other writers who empathized with the natural world, but the conservation movement at the time was driven by the desire to preserve natural resources for later use (in other words, so that humans could profit from or enjoy the indigenous animals' future existence). Many animals were destroyed because they were predators of, or somehow interfered with, the hunting objects of so called naturalists. The protection of game animals, combined with the complaints of ranchers that wild predators were harming their domestic herds and the creation and activities of new government programs such as the Bureau of the Biological Survey, the Animal Damage Control Act, and the National Parks Service, resulted in almost complete decimation of naturally occurring wild herds.[43] Sadly, all this occurred in the name of protecting land, simply so that humans could use the undeveloped United States frontier for work and play.

The 1920s saw the rapid growth and popularity of fashionable furs, followed by a backlash from activists who invoked images of materialism, the responsibility of maintaining ethical teachings for future generations, and even the juxtaposition of Christ on the cross with meek animals ensnared in the steel jaws of a trap.[44] A complete boycott of fur products never happened, but laws were created to make the trapping industry more humane.[45] Of course, the Great Depression resulted in a sharper focus on the sufferings of humans, but those fighting for animal rights never disappeared. The movement's dormancy was to be expected at the time, but a resurgence came about after the end of World War II, with a new focus: that of vivisection.

The term *vivisection*, which translates literally as "the cutting up of life,"[46] is often used to refer to medical experimentation on living animals. Like the animal rights movement as a whole, the fight against vivisection began in England, and like early animal rightists, anti-vivisectionists were ridiculed as anti-science and anti-progress, and thus against the overall advancement of human society. Any doubts the average person may have harbored about the value of scientific research and progress at the time were quelled by recently developed defenses against formerly lethal illnesses such as smallpox, rabies, and malaria, so animal rightists had to work hard to show that, despite such great scientific strides, animals' lives must still be respected. Anti-vivisectionist Henry Salt spoke for many others when he questioned the "moral hypocrisy of forming sentimental and intimate relationships with some animals while sacrificing others to science."[47]

When the American anti-vivisection movement first began in the late 1880s, women comprised the majority of the protestors. This may have been due in part to the empathy they felt with the animals being used against their will, similar to

Victorian women having little control over their bodies and fates in a male-dominated culture.[48] Arguments for vivisection were principally based on the presumption that medical advancements could bring about only good things for the world in general. Anti-vivisectionists tried to point out that the social ranking of doctors as intellectually elite implied that they had more responsibility for ethical treatments of other beings, including animals. They also brought up the point that whenever the suffering of animals is coldly ignored, it can lead indirectly to indifference to human suffering.[49]

Over time, as more and more research was done on animals, the differences between different species' responses to treatments was made clearer, so the argument that animal experimentation led directly to advances in human care held less and less water. Apparently it wasn't always useful to test things on animals, especially when it was proven that humans often reacted to the test substances or situations differently from animals. Using billboards, leaflets, and articles in national publications, anti-vivisectionists also tried to show the public what was occurring in the name of science. Distressing and often revolting images of test animals kept half alive, their bodies ravaged and missing limbs or organs, were commonly used for shock as well as educational value.

The medical and research communities bonded together in defense of vivisection and were often successful in swaying public opinion back in their favor. As early as 1884, the first committee to defend animal research was created by the American Medical Association.[50] Guidelines for laboratory animal care were developed and published in order to defend the system against activists' stories of inadequate care for research subjects. Such guidelines could certainly benefit countless animals involved, but there was still no governing body to police the new standards. Although attempts to establish an independent enforcing body to govern the use of animals in laboratories failed to pass the Senate, on the local level activists were able to pass laws in some states that prohibited vivisection in public and secondary schools.[51]

Vivisectors also wrote informative articles explaining the benefits of research on nonhuman subjects and disseminated these tracts throughout communities. They pledged stricter screening policies when hiring laboratory employees, which negatively affected the activists' practice of using insider information gleaned from converted or disaffected laboratory staff.[52] Terminology was modified to utilize words that were more scientifically based and less likely to remind readers of human or animal suffering.[53]

After World War II the economy in the United States was one of growth based in consumerism, partly propelled by scientific discovery that supported industrialization. While the resulting greater affluence detracted from issues relating to nature and ethical treatment of animals, the Civil Rights and Women's Liberation movements that arose in the 1960s and '70s paved the way for increased discussions of the legal rights of animals. Because people had more money and free time than before, instead of worrying only about the physical and financial survival of their families they could focus on "loftier" concerns such as ethics and morality. Many people moved from the cities to the suburbs and had more time to explore the world around them. Because former city dwellers were experiencing nature firsthand more than before, principles of ecology began to be held in higher esteem; but advances in science and manufacturing meant that the market was inundated with new chemical-based substances and products, most tied to animal testing in the laboratory.

It was during this period that activists were able to more successfully influence the government to pass laws that in one way or another protected the rights of animals. The Humane Slaughter Act, the Animal Welfare Act, the Endangered Species Act, and the Marine Mammal Protection Act were all created and put into effect at this time. Many more groups were formed and successfully promoted animal-friendly causes, the most popular of which dealt with laboratory and meat industries, as well as the extinction of specific animal species. Animal protection groups preferred moderate reforms over radical overthrows, either to avoid alienating the public or to avoid seeming too extreme in the post–Joseph McCarthy society.[54]

Over time, members moved away from the more moderate groups to form increasingly progressive organizations, and the more radical branch of animal activism expanded. Efforts to revise the methods used in modern meatpacking plants became the first major campaign of this generation of activists, who focused not on the danger to public health, as in the past, but on violations of the intrinsic rights of animals to avoid unnecessary pain and suffering.

The Federal Humane Slaughter Act went into effect in 1960 and required that all large-animal livestock receive some type of stun treatment so that they would not consciously experience pain before death.[55] This legislation applied to all federally contracted meat suppliers except those that operated under kosher rules, because of the religious necessity to uphold specific kosher rituals. Unfortunately for the animals involved, the kosher ruling that animals be conscious at the time of slaughter, combined with the Pure Food and Drug Act of 1906 ruling that no livestock could fall onto a bloody floor, meant that kosher animals had to be shackled and hung from machinery while awaiting their death.[56] Groups then fought to protect kosher animals from this seemingly unnecessary pain and suffering. Further dissension grew in regard to the separate kosher regulations, and activists became divided and scattered instead of presenting a unified front.

A major legal precedent was set in 1966 with the federal Animal Welfare Act (originally called the Federal Animal Welfare Act). This act won three major victories: It prohibited the theft of pets for medical research; it established standards of care for dogs, cats, primates, rabbits, hamsters, and guinea pigs in medical testing; and it authorized the Animal and Plant Health Inspection Service of the United States Department of Agriculture to research, license, and inspect facilities holding animals for research. Provenance records now had to be kept for all animals involved in research.[57]

The new law was not foolproof, for it did not apply to all animals used in medical testing (such as marine mammals), and it was unable to protect research subjects once the experimentation began, but it was a stepping stone towards establishing further regulations for animals living as research subjects. Additional regulation was needed to control the fast-growing research industry. It was now easier than ever to export exotic animals from their native lands via airplane, avoiding health problems like the suffocation and dehydration that had plagued the more antiquated travel methods. Additionally, sedatives allowed increasingly larger and less easily controlled animals, such as large primates, to be used in labs without fear of injury to researchers or damage to equipment.[58] As more and more animals, including primates, were transported into US labs for research, scientists perfected methods of keeping their subjects alive and breeding, supplying a constant stream of nameless, virtually identical subjects to slake the seemingly endless thirst of animal research.

Protecting animals during experiments came under the umbrella of later amendments of the Animal Welfare Act, which dictated minimum standards of care for animals while undergoing medical experimentation, required anesthesia be used whenever possible, and also held policing jurisdiction over other animal-care systems, such as in circuses and pet stores. The Animal Welfare Act eventually grew to protect and consider the natural social practices and mental health of certain research subjects, which was revolutionary in that it involved an empathetic, instead of merely investigative or coldly impersonal, view of animals in captive care.[59]

Researchers proposed a more compassionate animal research system that focused on the "three R's," which aimed toward "refinement of techniques to *reduce* potential suffering, toward *reduction* in the number of animals needed, and, where possible, toward *replacement* of animals by non-animal techniques."[60] [italics added] Through the pursuit of more moderate goals of alleviating pain and establishing standards of appropriate care for animals being experimented upon, progress in animal rights continued to be made without further internal divisions and without inciting the vigorous defensive propaganda techniques that painted animal activists as anti-science or as blocking the path of medical progress.

Engendering public sympathy through the media by way of photographs, catchwords, and stories that tugged heartstrings led to greater education and impacts on the industries of sport hunting and fur trading and the protection of endangered species. Images of clubbed baby seals, dolphins drowning in tuna nets, and whales that were becoming extinct in their native waters led to boycott campaigns, revisions of strategy, and successful protections for various animals under assault.[61]

The Endangered Species Act of 1973 was created in an effort to help curb one of the major side effects of rampant human expansion, the destruction of natural lands and resulting interruption of ecosystems.[62] It listed recovery methods and protections for species that were threatened or endangered. This was a pivotal law that gave specific rights to wild animals, rights that could not be supplanted by profit motives or other human interests. The Convention on International Trade in Endangered Species took place in 1974, and its related documents went a step further, specifically prohibiting trafficking in wild animals (as well as plants) in an effort to preserve threatened species.[63]

As American products were more highly scrutinized regarding their safety for use by the public, more and more product testing was conducted on animals. Everything from pharmaceuticals to cosmetics to cleaners to vehicles relied on animal testing before being released for sale or distribution to the public. This escalation in testing resulted in a number of backlash reactions as the public began to learn of various problems attendant to it.

For example, the use of baboons in automobile impact testing was especially disturbing to some people, due to the lack of care the primates would receive for the brain trauma they routinely suffered. A large, vocal, and persistent protest by animal rights groups resulted in Ford Motor Company agreeing to phase out their baboon tests and rely on non-animal alternatives.[64] Similar campaigns ended the United States Armed Forces testing of lethal gasses on dogs.[65] Such practices that harmed animals were reevaluated through the efforts of concerned citizens who were informed, thoughtful, and passionate about their beliefs. As Henry Bergh had begun it by policing the abuse of workhorses, the animal rights movement continued to grow as it remained dedicated to protecting creatures that were unable to defend themselves.

The 1980s was a highly charged time for protests against animal research, with escalating violence against research laboratories and people who worked in them. A 1981 undercover expose of appalling conditions for macaques being used in testing at The Institute for Behavioral Research began this new era of anger toward researchers and a greater determination to hold accountable those scientists who failed to comply with animal empathy standards. Two of the activists involved in the proceedings joined forces to create People for the Ethical Treatment of Animals (PETA), which quickly grew from fewer than 20 members to 400,000 members in the following ten years and continues to be a vibrant source of activism on behalf of animals in the United States.[66]

Currently PETA and their sister group, the Animal Liberation Front (ALF), are often discussed with some disdain on account of their customary fearlessness in confronting uncomfortable issues regarding the use and treatment of animals. ALF's tactics to draw attention to animal rights have ranged from vocal protests, violent threats, and raids of animal research facilities to thefts and destruction of personal property and what the FBI considers to be terrorist activities against animal researchers. In contrast, PETA focuses more on raising public awareness via education, media campaigns, and celebrity endorsements, and they remain one of the strongest forces protecting animals to date.

This chapter has laid out a somewhat broad overview of animal rights and its history in the United States. A more detailed view would likely become cumbersome in its complexity, as the movement has evolved over many decades and continues to do so. Such an outline as this can catalog many but not all of the changes made in the way American society treats its animals. In the end, it is clear that every small accomplishment of animal defenders has contributed to massive change over time, like flakes in a snowball, and what is in place today is a body of law, public opinion, and moral argument that strives to respect and protect all life forms on earth.

The history of animal rights has been provided as a backdrop to this book, as well as an inspiration, in the hope that from a panoramic view of the larger picture we may zoom in to look closer and reflect upon some of the details.

2. Abilities and Intelligence

We humans commonly react with astonishment upon discovering that chimpanzees can do something we consider special to humankind. Any evidence of intelligence overlap provokes the greatest skepticism, as the uniqueness of that quality in us is our most cherished illusion.[67]

— Geza Teleki

Modern biological and anthropological studies have created a clearly defined system of classifying primates. Almost all 230-odd species of primates share certain physical traits, such as pentadactyly (having five fingers and/or toes) and a clavicle, but there are four very distinctive physical characteristics that all primates display and that are not available in entirety on any animal that is not a primate: a bar of bone encircling or enclosing the eye sockets; nails instead of claws on most, if not all, digits; opposable thumbs; and the growth of the auditory bulla (the bone that encloses the inner ear) to the petrosal bone.[68] Primates are some of the slowest-growing and latest-to-mature members of mammalia. In addition, they are among the few types of mammals who keep their infants near them at all times.

Despite the many similarities among primates, there are also many differences that help to differentiate the various primate species. In this regard, it is helpful to consider the two main groupings of primates: prosimians and anthropoids.

The grouping commonly referred to as prosimian (mostly suborder *Strepsirrhini*) includes small, nocturnal mammals native to Asia and Africa, with pointed muzzles, wet noses, naked rhinaria (a patch of bare skin around the nose), teeth that form a toothcomb (for all species except the aye-aye) and claws on at least some of their extremities. This grouping includes the following species: tarsiers, lemurs, sifakas, indris, aye-ayes, lorises, pottos, and bush babies [galagos].[69] Prosimians tend to be

47

solitary, but because their home ranges often overlap, there can be communication between individuals through vocalizations and scent-marking via urine and feces.[70]

The remaining primate species are referred to as anthropoids, which include both New World and Old World primates. All New World primates are monkeys (of a parvorder called *Platyrrhini*) that live in Central and South America and have a wider septum and an additional bicuspid in their dental pattern.[71] Examples of New World monkeys include marmosets, tamarins, sakis, uacaris, howler, spider, capuchin, woolly, squirrel, night, and titi monkeys. These primates are arboreal and territorial and, like prosimians, communicate their presence partially by scent-marking their surroundings. One family of New World monkeys, *Cebidae* (which includes capuchin monkeys), has a long, prehensile tail with a bare patch of skin that is not only used in brachiation, to swing on branches, but can also be used as a limb to grab things within reach.[72] Most New World anthropoids are not sexually dimorphic (meaning that there is not much physical differentiation between males and females).

Old World primates (parvorder *Catarrhini*) are principally from Asia and Africa and can be distinguished by their downward-facing nostrils and flat fingernails and toenails. All Old World monkeys have ischial callosities, which are calloused pads on their hindquarters, on which they sit,[73] and they show a marked sexual dimorphism. In addition to humans, primates in this group include apes such as chimpanzees, bonobos, gorillas, orangutans, and gibbons, as well as monkeys such as macaques, baboons, vervets, mangabeys, guenons, and colobus and patas monkeys.[74] Old World nonhuman primates are skilled at manipulating objects and can be trained to perform various tasks. Most of these primates can subsist on quite a varied diet, so they thrive when living near human populations, sometimes to the point at which local human cultures consider them pests.

From the smaller, nocturnal, arboreal prosimians, such as the lemurs and lorises of Madagascar, to the powerfully built, sexually dimorphic baboons of the African savannah, the physical characteristics and behaviors of primates can run the gamut. The smallest primate, the mouse lemur, can weigh as little as a few grams, while the mountain gorilla can weigh up to four hundred pounds.[75] Primates can be nocturnal (active at night), like the tarsier, or diurnal (active during the day), like humans. Some are even cathemeral, or active only sporadically during a twenty-four hour period and otherwise at rest both day and night, such as the lemur.

Primate social groups can take many forms: monogamous pairings, like indris, siamangs, and titi monkeys; one male living with several females, such as the colobus monkeys, gelada and hamadryas baboons; or multi-male and multi-female groupings, which are the case with most New World monkeys and many Old World monkeys, as well as the African apes. An individual's dispersal from the group upon puberty can vary, depending on the species. Some social groups are female dominated, like the ring-tailed lemurs and the bonobos; some are male dominated, such as Western lowland gorillas and baboons; and some primates don't even have social organization in their natural habitats, such as the mostly solitary orangutans of Borneo and Sumatra. However, the species that are male dominated always reflect some sort of sexual dimorphism. For example, baboon males dominate their societies, and are almost twice the size of females, with large canine teeth that they flash in frightening and threatening displays of power.

Some primate species are completely arboreal, such as colobus monkeys, and some are terrestrial, like gorillas. Others live both in and out of the trees, like chimpanzees. In general, the larger a primate is, the more likely he is to spend time on the ground, simply on account of the basic rules of physics, which may cause him to fall from fragile tree branches. This is not necessarily species specific, as is the case with orangutans, which will live fully arboreal lives unless an individual gets too heavy (generally, only males tend to grow that large). A larger individual will increase its time on the ground out of necessity and self-preservation.

Some actions and behaviors appear to be confined to certain families or groups of similar species. For instance, monkeys do not like eye contact, but great apes do. Although direct eye contact serves as a reconciliation and social tool among apes, monkeys avoid staring at others, since to them it indicates a threat.[76] This leads anthropologists to believe that the common ancestor that humans share with chimpanzees must have been reliant on eye contact as a social cue to direct behavior.[77]

Apes reflect displeasure by frowning, as a universal gesture with the brows furrowed.[78] Additionally, apes all beg in the same way: With palms up, they will reach out to another individual as if asking them for a favor or to implore them to follow or join in an activity. Such manual gestures have only been observed in apes and humans. Scientists believe that these instinctive gestures may have constituted the first language developed by early humans, a sort of proto–sign language that required voluntary control and, later on, allowed for more complex communications.[79]

Infants of all nonhuman primates cling to the mother with a strong, instinctive grip that human babies do not exhibit. They remain in the mother's care for the first few years of life, varying with the species. Their extreme proximity to their mothers means that they very rarely cry,[80] while human babies cry to get their mothers' attention because at many points throughout the day they may not be in their mother's embrace. Nonhuman primate babies are almost always found hanging from the front or back of their mothers' bodies and, thus, have little need to cry out.

The closeness of the mother-infant relationship among nonhuman primates ensures that a mother's offspring will develop into a healthy, well-adjusted individual. If this bond is interrupted, due to the death of the mother or outside forces, such as human involvement, the infant suffers terrible physical and emotional side effects. Infants who were not able to enjoy a normal bond with their mothers often grow up to reject their own infants, as well.

Female primates of all species appear captivated by new infants born within their social groups, regardless of their biological connection to a particular youngster. Although various species have different customs regarding the touching of an infant not their own, females of almost all nonhuman primate species will gather around a mother and her youngster, grunting and observing as the infant moves about and learns new behaviors.[81] The mother-child bond can persist even after death, illustrating the strength of the connection between parent and offspring. For example, the daughter of an elderly chimpanzee who was dying at a safari park in the United Kingdom held a vigil over her mother's body. At a research site in Guinea, two different chimpanzee mothers were observed carrying and tending to the mummified corpses of their infants for up to ten weeks, including carefully shooing flies away from the bodies of their deceased offspring.[82]

Historical evidence illustrates many similarities and also differences between humans and the other primate species. Although primate fossils have been found dating from five million years ago, there is not yet universally accepted agreement about the moment when humans and other primates split to go their genealogically different ways. Estimates of this date vary from six to eight million years ago.[83] Hypotheses for what caused the split vary, as well, from diet (most primates are mainly vegetarian, but some nonhuman primates eat meat on occasion, such as chimpanzees, bonobos, and capuchins) to bipedalism (walking upright on two legs) to other evolutionary occurrences that potentially freed up hands and allowed increased caloric input to fuel cranial development. Whatever the cause, it's important to note that the general consensus is that the split from our common ancestor, and between what we now know as humans and other great apes, occurred gradually, over time, like everything else in evolutionary history.

Divergence from this shared ancestor split traits of human and nonhuman primates and eased the transition out of the forests. Bones discovered by anthropologists Ronald Clarke and Phillip Tobias revealed a partial left foot of an *Australopithecus* ancestor living at least three million years ago that proved its owner was capable of bipedalism and also had a grasping, simian toe that would have allowed for arboreal locomotion.[84] Anatomist Randall Susman believes that similar evidence of transitional locomotive patterns can be seen in the knuckle-walking of chimpanzees and gorillas, which allows quadrumanous locomotion (four-handed walking) but also can temporarily free up the hands for food or object procurement and even for tool making and use.[85]

It was Charles Darwin who claimed that membership in the *Hominidae* family was marked by bipedalism (upright posture), tool use, and a higher level of intelligence. Bipedalism allowed early humans to use their hands to make tools, and in order to create tools the brain had to be somewhat clever and creative. But it was later proven that early hominids were walking upright for more than two million years before tool use first appeared, and the increase in brain size had occurred more than four million years before that! The transition from "animal" to human, it appears, cannot be so easily explained.

Other opinions on the cause of bipedalism included the idea that standing upright made the individual appear more dominant or that it allowed increased visibility of predators over the natural environment's grasses and other plants, and so, through natural selection, the genes continued on. Some experts have also argued that bipedalism was less calorically expensive than quadrumanous locomotion, or that it reduced body exposure to the sun, both of which permitted an easier life for individuals with the trait. Upright posture would also facilitate tree climbing. Conversely, perhaps skilled tree climbers developed a more upright posture gradually, which then evolved into upright locomotion on the ground. Whatever the origin(s) of upright posture, a bipedal individual was able to take advantage of his freed hands and more easily manipulate and use tools, which could aid his dominance over predators and, ultimately, his survival.

As similar as nonhuman primates may be to humans, just as interesting are the unique traits each separate species has developed in response to survival in its specific environment. For example, chimpanzees continued to evolve for five million years after diverging from the common ancestry shared with humans. (Humans continued to evolve, as well, albeit in different ways than did chimpanzees.) Therefore, while humans are typically considered the more highly evolved species, chimpanzees are also highly evolved, although their unique traits evolved differently from those of humans. Humans may have developed a more sophisticated language, for example, but chimpanzees developed a more sophisticated method of arboreal nesting, perhaps equally important to their survival. This begs the question, why is one highly evolved trait, such as language, considered superior to another, such as arboreal skill? Is it simply because humans have the capacity for language and thus proclaim this trait to be of greater importance?

An activity for which nonhuman primates are well known is reciprocal grooming—searching through another individual's fur to pick out nearly invisible foreign matter. This activity may be shared between two individuals or perhaps among an entire group of primates. Grooming is an evolutionarily developed trait that is integral to the health of these primates, as it results in the removal of parasites and other unhealthful invaders that may be hidden in the fur. Of course, humans groom their children, but they don't typically groom each other as a form of social interaction.

Grooming is also very important social tool for nonhuman primates. It not only soothes and interconnects the group as a whole and reinforces existing hierarchies that define the group, as subordinates groom the more dominant members, but is also used to calm a member who may have been upset or offended during an earlier exchange. This "reconciliation hypothesis," as defined by bonobo expert Frans de Waal, "predicts that individuals try to 'undo' the social damage inflicted by aggression, hence, they will actively seek contact, specifically with former opponents.... Reconciliation ensures the continuation of cooperation among parties with partially conflicting interests."[86] Grooming, then, can be thought of as an olive branch, a way to atone for past sins, or it can be a calming activity between two primates who simply like each other's company and enjoy making each other feel good.

Primates are, in general, very social animals. Most species live in groups, despite the possible pitfalls of group living, which can include increased food competition and a higher likelihood of intra- and intergroup aggression. Species-specific patterns of males or females leaving their natal groups upon sexual maturity help to ensure genetic diversity within a group, as well as aiding in the genetic continuation of the individual.

It seems that one of the benefits of group living is purely mental and/or emotional: primates derive comfort from each other. Whether via grooming, physical touching, patting, hugging, or even kissing, the brains of most primates appear wired to seek out physical companionship, even during seemingly insignificant times of play. This compulsion suggests that their brains are capable of keeping track of the social rules of their groups, such as the dominance hierarchies that always exist, various warning calls, and methods of responding to an outsider who appears to the group. Some scientists have speculated that great ape and human brains evolved to their high capacities in response to the need to keep track of complex social cues and rules.[87]

The social networking capabilities of primate brains vary by species. Some primates, especially chimpanzees, are quickly able to recognize individuals from their past. This was evident when Boee, a chimpanzee who had been taught sign language by primatologist Roger Fouts in the 1970s, saw Fouts for the first time in 16 years and recognized him immediately, signing his name excitedly.[88] New World monkeys also have long-term memories of individuals and can recognize significant individuals from their past by sight, scent, and even, in the case of humans, the sound of their footsteps.[89]

This mental sophistication is required to remember group history and alliances, even when pertaining to group members who have passed away. Chimpanzees in captivity have expressed behaviors that are comparable to human mourning. If a group member appears near death, others will stand vigil nearby, with increased grooming and observation practiced. Once a death has occurs, group members will caress and spend time near the deceased, and emotional behaviors of the typically expressive apes will be noticeably subdued overall. Earlier in this chapter it was mentioned that chimpanzee mothers whose infants have died were observed carrying around the mummified corpses for up to ten weeks. These actions in the face of death have been compared with similar denial behaviors in humans facing loss and coping with extreme grief.

Primate sociality, for most species, is also fluid: not every member of a particular species has the identical social patterns of its same-species peers. This is likely the reason that dominance hierarchies preside over so many primate communities. Just as with human beings, some nonhuman primates have greater aptitude in particular activities than others, and the value of specific strengths may affect both an individual's worth in a given society and his survival in the wild.

Mental abilities vary widely among primate species. Advanced mental capabilities permitted humans to differentiate themselves clearly from the rest of the primate world. Because primates resemble each other in so many ways, researchers commonly study primate intelligence and abilities in search of surprises amidst the folds and curves of the brain. Cognitive ethologists, scientists who study the processes of animal intelligence, often focus on the multifaceted minds of great apes because they offer the greatest range of abilities and because there has already been much prior research on those species. Great apes such as chimpanzees are relatively easy for laboratories to procure, and as they are biologically closer to humans than are monkeys, this research has wider application to human culture and greater worth to scientific-grant-awarding institutions.

In the 1920s, early research by psychologist Wolfgang Köhler tested the problem-solving abilities of chimpanzees as they attempted to obtain bananas hung above them. When Köhler described their capabilities as evidence of foresight, the anthropomorphism alarms went off, as people were shocked when asked to consider that animals could display such an elevated level of thought.[90] (Anthropomorphism, the assessment of human characteristics, abilities, and behaviors in nonhuman beings and a consequent assumption of attendant human feelings, thoughts, and motives, is often hurled at biologists as an insult. Generally considered to be anti-scientific and unprofessional, anthropomorphist may see human emotion and reasoning where no more than biological or instinctual cause-and-effect exists. This inclination will be discussed at greater length throughout subsequent chapters.) The threat of anthropomorphism didn't fully discredit Köhler's work; however, while his findings gained a greater following, many scientists appeared reluctant to leave behind the accepted belief that their research subjects were more than a bundle of nerves and muscles. For example, physiologist Ivan Pavlov (of the "Pavlov's dogs" operant conditioning research) deemed Köhler's conclusion "disgusting."[91]

A prominent figure in the early days of primatology was Nadezhda Ladygina-Kohts, a Russian researcher who studied chimpanzee cognition in the 1930s. She deduced that ape empathy was stronger than monkey empathy (a conclusion that has been supported by additional research in more recent years). Ladygina-Kohts described what would happen every time she pretended to cry: This behavior caused a young chimpanzee subject to stop what he was doing immediately, approach her, visually examine her face, and attempt to make her feel better with light touches.[92] This type of consolation behavior is common among great apes, as is grooming, and it is compelling to note that when it occurs, the agent does not gain anything tangible from the consoling behavior. Evolutionarily speaking, this behavior is confusing, for why would a wild animal engage in conduct that does not directly improve his chances for survival?

Further research by primatologist Frans de Waal has shown that the advanced mental capabilities of apes permit them to understand the viewpoint of an outsider (and to a greater degree than the abilities of a monkey allow).[93] Numerous other examples exist of chimpanzees living in captivity that intuit the desires of other group members and altruistically help them achieve their goals. This impulse can even cross the species barrier: In 1996 a gorilla named Binti-Jua gently carried to safety a four-year-old boy who had fallen into her enclosure at the

Brookfield Zoo in Illinois, fending off other gorillas with a threatening growl.[94]

Although the higher intelligence of such social animals seems to be related to their involvement in a community, it's important to note that less social primates, such as orangutans, do not appear to suffer intellectually for lack of interaction. Much in the way that chimpanzee intelligence has been compared and contrasted with human intelligence, orangutan thought processes have been studied as well. Anne Russon, a primatologist and professor at Toronto's York University, studies the orangutans of the Nyaru Menteng sanctuary in Borneo and compares their intelligence to that of a three-year-old human, but with a caveat: "They don't have a child's mind."[95] According to Russon (who has been observing the same group of orangutans every summer for more than thirty years), the slower-moving orangutans are not slower intellectually; they appear to be constantly storing their experiences for future reference.

Because of scarcity of food in their home ranges, Bornean orangutans live and travel alone (except when a mother has a youngster, as it takes years to raise the infant to a state of independence), although there may be temporary groupings at a particularly dense feeding spot. The slightly smaller Sumatran orangutans, now recognized as a species distinct from orangutans in Borneo, have less sparse feeding spots and thus are more likely to form social groups around meals (groups that tend to consist of numerous females with one dominant male for protection).[96]

Borneo orangutans have developed their own unique set of smarts that benefit them in ways appropriate to their lifestyle. For instance, although they have no need to remember social alliances and hierarchies, evidence shows that they can recall the fruiting patterns of trees that they visit for food only once every eight years. Additionally, their seemingly innate ability to mimic detailed physical tasks has become a topic of great interest to scientists studying the human mind, as their level of observational skill and applied creativity is thought to be what also facilitates education and comprehension in human beings. Orangutans can be seen observing actions and then seem to break down the separate movements in their minds, and only when they have figured out a sort of internal plan of attack do they attempt to recreate the action they originally observed.[97] Known to be thoughtful and precise in captivity, orangutans will often act out an internal script in order to deceive those around them into giving them something they desire, usually access to something clandestine via a glitch that had gone unnoticed by the human eye.

The impulsive, emotional outbursts of young human children, which also seem so recognizable in chimpanzees, have no place in orangutan life. They have been found to point at items with their eyes, rather than with their hands, and do much communicating with subtle eye and body movements.[98] Detail oriented and clean, they tend to prefer order and patterns, evidenced by the way they will often arrange sticks and leaves in particular configurations around their nests and living areas, for no apparent reason other than decoration or to satisfy some internal need for order in their jungle home.[99]

Precision rules orangutan society. Their relationships, like their fruit trees, have been described as being dispersed over time and space. Says primatologist Carel van Schaik, "[They] don't have to meet every day. There is a lot more structure than meets the eye."[100] The subtle social networks of orangutans are the antithesis of the explosive, dramatic relationships of chimpanzees, but that does not imply that one is better or more advanced than the other. Rather, they have developed in response to very specific environmental needs.

Primates as a whole have a relatively larger ratio of brain size to body size, when compared to most other mammals.[101] The frontal lobe of the brain's cortex (the neocortex) is the center of many thought processes—including creative thought, decision making, memory, and emotion—and is almost the same size in chimpanzees and humans.[102] Psychologist Steven Walker admitted that "much work has been done since Huxley emphasized 100 years ago that 'every principle gyrus and sulcus of a chimpanzee brain is clearly represented in that of a man,' but there is nothing that contradicts his conclusion that the differences between the human and chimpanzee brains are remarkably minor by evolutionary standards."[103] Evidence of this similarity is found throughout the field of cognitive research.

Developmental psychologist Michael Tomasello and his colleagues at the Max Planck Institute for Evolutionary Anthropology tested the learning abilities of chimpanzees versus those of small children. When a child was shown a simple task, the child would mimic the trainer to achieve the proper result. This process of imitation is one of the most integral ways that humans learn to survive to adulthood.

On the other hand, chimpanzees, when shown how to complete the same simple task, would often go about it in a different way; they would achieve the same result but through a unique process, which they determined for themselves. Tomasello called this emulation (as opposed to imitation) and inferred that the act of imitation is uniquely human and that the inability to mimic the steps (and to understand the reasons to mimic the steps) to achieve a specific goal is something that separates other species from humans.[104]

Although orangutans can easily imitate typically human tasks, primatologist Biruté Galdikas has an interesting theory for why this occurs. She thinks that they do this as a social mechanism to bond with whichever human first exhibited the task. This sort of copycat behavior is a way to equalize two individuals and to show a relationship between them that otherwise might be prevented from expression due to a communication barrier.[105]

Cognitive mapping, the ability to form mental representations of items that might be hidden in various locations, usually occurs in human children prior to turning three years old. Evidence in captivity and in the wild shows that adult chimpanzees and bonobos have this ability, as well,[106] and capuchins and gorillas have been shown to pass tests involving concepts such as object permanence (the ability to know an object is present even if hidden from sight).[107] Additional tests have proven that some nonhuman primates are capable of recognizing mirror images and rotations of symbols, as well as the relationships between scale models and full-size objects,[108] and of sorting, comparing, and classifying objects.[109] Research by Sally Boysen through the Primate Cognition Project showed that language-trained chimpanzees may be able to comprehend what is called a second-order relationship (the more abstract relationships among seemingly very similar objects, such as metal nails and screws, or fresh apples and oranges) because the language itself gives them an additional level at which to organize their thoughts about the world around them.[110]

Further evidence of higher thought among great ape species can be found once again in chimpanzees and orangutans, who use a simplistic barter system in the wild. High value is placed both on sexual access to ovulating females and on meat, and at times one may be exchanged for the other. This is evidence of higher thought processes such as representation (x amount of meat is equal to copulation with female x), reciprocation (if I give you meat, I deserve sexual access in return), and alternative planning for future options (if I don't eat this meat, I can give it to this female in return for sex), among others. [111]

A Japanese study by the researcher Tetsuro Matsuzawa, featuring a chimpanzee named Ai and a computer math program, showed that not only could Ai distinguish distinct numbers she viewed on a screen, she could also understand the relationships between numbers. She was easily able to select randomized numbers in ascending order, accurately and without prompting from the researchers. Ai's son Ayumu has regularly completed numerical tasks with even more accuracy than Ai, and researchers believe this is due to greater eidetic memory in youngsters—that they are better at making accurate mental images of complicated puzzles.[112] Not only are young chimpanzees better than older chimpanzees at such tests, but they outperform adult human beings, as well. In testing on color recognition, not only can chimpanzees differentiate various colors, they can match them accurately with their correct Japanese symbols.[113]

Other experiments have shown that chimpanzees can understand the relationships between numbers and also add fractions, and they can even understand more advanced concepts such as conservation (the idea that a piece of clay possesses the same mass regardless of its shape).[114] Chimpanzees also display an understanding of reciprocity: It was found that an individual who had received grooming from another individual was more likely than usual to share food later in the day with the groomer. Although this behavior could be explained using the theory that the groomed individual was in a better mood after receiving so much attention and was thus more likely to share food in general, it appears that groomees tend to share food specifically with their groomers after a session. This behavior requires abilities involving memory and the conception of gratitude, which may not be as well developed in other primates.[115]

Although great apes such as chimpanzees may have mental abilities greater than many other primate species, it's not only the higher primates who are able to comprehend basic mathematics. Marc Hauser of Harvard University proved that rhesus macaques were able to add one plus one and, when presented with two items, were also aware if one was subsequently removed.[116] These results and others like them prove, at a minimum, that some primate species show greater comprehension of number relationships than do very young human children.

When reviewing accounts of the intelligence of certain primate groups, it's not uncommon to encounter a related discussion of their culture, be it pertaining to child-rearing practices or location-specific methods of cracking open nuts. What is culture, exactly? Primatologist Carel van Schaik defines it as socially transmitted behavior that is customary or habitual (exhibited by most members of a group, or at least most of the relevant members of a group) in one location but is absent from another group at another location and cannot be explained genetically.[117]

Evidence of culture can be found in the following categories: labels (ways of recognizing objects like food or predators), signals (socially transmitted variations of displays), skills (such as tool use), and symbols (variations of signals that have become unique to a social unit or population).[118] Van Schaik believes that although chimpanzees and orangutans show cultural evidence with labels, signals, and skills, only human beings have continued on to show evidence of symbol use in cultures, due to humanity's advanced communication and education systems. This definition of culture seems to be generally accepted in the field of anthropology.

Primatologist William McGrew of Miami University in Ohio has created a guideline consisting of specific behaviors that groups must exhibit, which qualify as evidence of culture:

1. Innovation (A new pattern is invented.)
2. Dissemination (The pattern spreads to other individuals.)
3. Standardization (The pattern is consistent among individuals.)
4. Durability (The pattern is performed even when others are not around.)
5. Tradition (The pattern is transferred over generations.)
6. Diffusion (The pattern spreads to other groups.)

While evidence of each step has been observed in various primate studies, as of yet there has been no nonhuman population to meet all six requirements with one single action or pattern.[119] Nonetheless, it's important to note that some human populations might also fail to meet all six criteria definitively, and McGrew believes that a more basic interpretation of culture is still acceptable in such cases.

Many cultural anthropologists would argue that culture is, by definition, a purely human attribute. Culture, they say, uses symbols, and the behavior of animals is not symbolic, thus nonhuman beings are unable to have culture. Perhaps the disagreement between cultural anthropologists and primatologists can be resolved if semantics are taken into account. The word *culture* was originally developed by humans and for humans; thus, it may be fundamentally impossible to claim that nonhumans have cultures, per se. Of course this use is purely literal, and there seems to be a good deal of evidence of nonhumans exhibiting their own versions of what we consider to be cultures.

Anthropologist and philosopher Barbara Noske analyzed the habit of anthropologists (by definition, scholars of humans) to ignore any signs of culture, community, or general intelligence in animal communities. She observed that even though the characteristics listed above may be very obviously present in nonhumans, anthropologists don't look for them because they're operating under the presumption that only humans have cultures and enriched social communities. She concluded, "If one preconceives humans to be the sole beings capable of creating society, culture or language, one will thereby have pre-empted 'ape' forms of society, 'ape' culture and 'ape' language almost by definition."[120] That assumption is a risky business that could result in the ignorance of possible cultures in other species.

Primatologists have discovered, for example, unique methods of tool use that have developed in geographically separate communities of chimpanzees. East African chimps have created tools to fish through termite and insect mounds, whereas West African chimps have created hammers to crack the hard shells of nuts. Although both coasts of Africa share the same food resources of insects and nuts, the groups of chimpanzees living on each side developed different methods of obtaining that food. As these methods were passed down from one generation to the next, the behavior stuck.

Explaining away such unique location-specific behavior as simple "behavioral variation" indirectly reinforces a stereotype of nonhumans as acting purely on instinct and without any particular intentions or forethought; but the fact that the environments in which these two behaviorally diverse chimpanzee populations live are similar may instead lead to the conclusion that environmental factors are likely not a hidden cause of what appears to be cultural, and the same finding holds true for orangutan behavior differences.[121] Many scientists are quite content to label such occurrences as evidence of culture. For instance, in a 1999 article in *Nature,* leading chimpanzee experts including Jane Goodall, William McGrew, Toshisada Nishida, Richard Wrangham, and Christophe Boesche found among their assorted research more than three dozen instances of cultural variations among social groups.[122]

Reasons have been developed to explain the geographical differences in chimpanzee tool use (and refute the existence of chimpanzee culture), such as the proposal that genetic differences result in the development of varying food procurement traits, or that different environments simply make it easier for chimps in one area to use sticks instead of rocks. Such theories are often refuted, because chimpanzee groups are genetically identical in broad terms., and in similar environments, chimpanzees will seek out the specific tools their communities favor. Rocks to use as hammers are not more easily found in West Africa, but the chimps that live there seek them out, and chimps in East Africa do not.

There are many other examples of such behavior in nonhuman primates. Capuchin monkeys have been observed not only using hammers and anvils to crush found nuts, but also purposely transporting selected hammers to proper anvil sites, showing evidence of forethought and planning.[123] On the island of Koshima in Japan, macaque communities will wash sweet potatoes both in a freshwater stream and in the salty ocean before consumption, something that has been dubbed "seasoning behavior" by the observing primatologists.[124]

Chimpanzees in Fongoli, Senegal, fashion and use tools not only to catch termites, as Jane Goodall observed, but also as spears with which to hunt meat (specifically bush babies hidden in hollow tree trunks), the first instance of nonhumans creating and using deadly weapons.[125] Chimpanzees at Goodall's research site at Gombe have been observed selecting and carrying termite-fishing tools even when not near a termite mound,[126] proving the depth and scope of their abilities of mental representation and planning for future activities. They have also been observed seeming to cooperate during group colobus monkey hunts and, in a study at the Max Planck Institute for Evolutionary

Anthropology in Leipzig, Germany, using water as a tool to float out-of-reach peanuts within grasping distance of their hands.[127]

Chimpanzees, orangutans, and gorillas have demonstrated self-awareness in mirror tests,[128] something that no other species but humans has thus far exhibited. This awareness is equivalent to that of a 2-year-old human child. Macaques, gibbons, and baboons are able to recognize the existence of an animal in a mirror reflection but so far have not exhibited behavior showing that they realize it is themselves being reflected.[129] In these cases, although pygmy marmosets and cotton-top tamarins have shown precursory evidence of self-recognition in mirrors,[130] most other primates tend to search behind the mirror, looking for an animal that they believe is staring at them through the glass.[131] Psychologist Gordon Gallup, head of the first mirror self-recognition tests of nonhuman primates, in the 1960s, wrote that "these data would seem to qualify as the first experimental demonstration of a self-concept in a subhuman form."[132]

It's important to note that tool use among nonhuman primates does not exist solely to find or prepare food. Orangutans will swing on flexible tree branches and use leaves as napkins or towels to clean their bodies.[133] Chimpanzees sometimes use leaves as cups to trap, collect, and pour rainwater. They also use branches and leaves to cover themselves when it rains. They may use small sticks to protect feet when climbing prickly trees, and also in nostrils, as humans would use a Q-tip in a stuffy nose. Female chimpanzees will even play and cuddle with sticks, acting as if they were infants, in the way that human children play with dolls.[134]

Bonobos will drag sticks through the dirt, signifying to the group that it's time to move on to a new resting site. Capuchins use leaves to sponge up water and have been observed beating deadly snakes with sticks. Macaques have been observed soaking in natural hot springs and throwing snowballs for entertainment in the winter. The significance of findings such as these illustrate that the inventiveness of primates is not limited solely to food but also to comfort, pain avoidance, protection, and self-medication.

Music appreciation and creation is one of the cornerstones of human culture, and evidence increasingly supports this trait in apes as proof of yet another way they exhibit culture. Great apes in the wild and in captivity have been found to be very receptive to music. When a musician played a guitar in various ape sanctuaries, he found that the chimpanzees and orangutans were captivated by the music and were enthralled with the guitar producing the sounds. They participated in the performance by bobbing their heads and softly hooting along, which sounded to him like an attempt to harmonize with the song. Musicians Peter Gabriel and Paul McCartney have played music with bonobos, who supposedly have a good handle on rhythm and pitch. Some sanctuaries will play recorded music as a surefire way to calm their primate inhabitants when they are especially riled. Likewise, chimpanzee groups in the wild have been observed moving rhythmically and vocalizing at times of high emotion, such as during thunderstorms and grass fires.[135]

There are many clear, documented cases of traditional behavior passed through members of primate societies, often surviving after generations. Craig Stanford of the University of Southern California believes this is evidence of the existence of nonhuman primate cultures.[136] Each new group that adopts a particular behavior also adapts it. Often started by younger members of a group and then traveling to the older members, it moves from individual to individual, morphing and meandering through a community. Conversely, some activities are believed to be taught by elder members to the younger members of a group, particularly if the behavior involved is detailed or exacting, requiring practice and learned expertise to perfect it.

The fact that certain behaviors are learned via nurture instead of nature, ingrained and not instinctual, is often presumed to be evidence of primate culture. Compared with other species, primates happen to be good at imitating others, a skill that may appear basic on the surface but which is actually very complex, involving planning, consideration of cause and effect, and the physical dexterity to recreate the behaviors observed. It may or may not be influenced by environment.

The similarities between human and nonhuman primates include not only intelligence and occasional benevolence; there are additional behavioral characteristics shared between primate species that are not things to celebrate. One such trait occurs in chimpanzee society, where sexual dominance and beating of females is common (something that, unfortunately, also happens all too frequently among humans).

Some nonhuman primate species practice infanticide, including monkeys, chimpanzees, and gorillas. For example, when a dominant male langur monkey takes control over an already existent group of females, he systematically kills all the infants. This morbid practice is actually supported by evolution, for the infants' deaths allow the mothers to be fertile sooner than they would have been, so the dominant male can impregnate them sooner and ensure continuation of his genes. Additionally, there will be no youth in the group to compete with his own progeny for dominance. The numbers of infants killed in such attacks are relatively high: 35% in grey langur monkeys, 37% in mountain gorillas, 43% in red howler monkeys, and 29% in blue monkeys.[137]

In 1925, when Raymond Dart discovered *Australopithecus africanus*, the ancient human-ape ancestor who proved Charles Darwin's evolutionary theories correct and provided a physical record of what early man was like, Dart postulated that this being was a carnivore who ate his prey alive, violently tearing up the carcasses and drinking the blood. Scientific theories such as Dart's helped encourage the bloodthirsty ape imagery in popular culture and also encouraged humans to ascribe their war and violent tendencies to some link from their not so distant past.[138] Dart's theories were later proved incorrect, as evidence from examination of *Australopithecus* skulls revealed injuries not from man-to-man combat but from larger animals, proving that early man was easy prey.[139]

It's interesting to consider that if Dart's theories had more quickly been disproved, nonhuman primates might have been seen through more compassionate and less fearful eyes. Ethologist Karl Lorenz noted that although most other animals in the world do not participate in interspecies warfare and genocide, humans do.[140] It's as if humanity's strength and violence evolved faster than its system of Darwinian checks and balances.

On the other side of humanity's affiliation with the animal world, consider the bonobo, the species to which humans are most closely related and which separated from us genealogically only six million years ago. These typically peaceable beings live in a matrilineal society ruled by cooperation and social comforting, where sexual couplings placate disputes and where dominance and cannibalism don't exist. In fact, "the use of sex to promote sharing, to negotiate favors, to smooth ruffled feathers, and to make up after fights is enough to make it the magic key to bonobo society."[141]

For bonobos, sex is a recreational and social pleasure, not used just to assert dominance and conceive offspring but also serving the function of allaying competitive aggression and calming excited nerves. It's an integral part of bonobo daily life, more so than with any other primate species, and can occur in unconventional pairings, such as homosexual and intergenerational, and in numerous positions as well.[142]

Bonobo peacefulness even extends across the species barrier. Whereas chimpanzees will hunt and eat monkeys in the wild, bonobos have been observed catching monkeys and keeping them as playthings for their own entertainment. They will groom and swing the smaller animals like toys and even mount them sexually, acting confused and playing a little rough when the monkeys don't cooperate as expected.[143] Their diet does include a small amount of animal protein (mainly insects, and rarely small rodents and duikers), although studies have shown that 99 percent of their protein is plant based.[144] All this being said, primatologist and bonobo expert Frans de Waal notes that it's important not to romanticize bonobos as an ideally peaceful relative: "Even if strikingly pacific, they are not the long-lost noble savages. All animals are competitive by nature and cooperate only under specific circumstances and specific reasons, not because of a desire to be nice to one another…bonobo society is not all rosy. The species is no exception to the rule that cooperative tendencies are best understood in conjunction with competitive ones, even though I agree that in bonobos the emphasis seems to have shifted to the former."[145]

Bonobos, first referred to as pygmy chimpanzees, were not recognized as a separate species until 1929.[146] Their slimmer frames, more upright posture, and less furry bodies illustrate how much closer relations they are to humans than are chimpanzees. Had they been noticed and studied earlier, perhaps their genetic proximity to humanity might have influenced human culture and attitudes more than traits of other apes that were observed or assumed. Sexual expression, sexual equality, and social networks might have been seen more as activities preferable to war, territoriality, and other aggressive mainstays of human culture.

It's been proposed that perhaps the bonobo's natural reliance on pleasure and bonding through physical touch could begin to convince humans that such traits are innate in us, as well, and perhaps ought to be considered more praiseworthy in the modern world. After all, if morality is based at least in part on what is natural behavior for us, perhaps humans could be more accepting of the physical liberties displayed by our closest relatives.

The various nonhuman primate species' differences and similarities to humankind has given rise to innumerable studies that research not just what primates do, but *how* they do certain things. After years of field studies of wild chimpanzees in Bossou, Guinea, primatologist Tetsuro Matsuzawa believes that one of the key differences between human beings and other primates lies in how skills are passed down from one generation to another. Primates, he claims, do not teach (despite research by other primatologists that seems to prove otherwise). Youngsters may watch their parents perform an action, he says, but when the youngsters try to complete that same action themselves, they must rely on trial and error to guide them, not the correction or encouragement of a parent.[147] Evidence discovered by Christophe Boesch of chimpanzee parents possibly teaching their infants how to crack nuts with an anvil has been rare—only twice in more than 70 hours of observation[148]—lending credence to

Matsuzawa's theory of primate societies being void of true education.

The variety of theories presented in this chapter reveals just how difficult it is for humans to know what drives nonhuman behaviors. Some of the most well-known biological theories derive from Charles Darwin's work on evolution, and although his popularity has helped promote protections of nonhuman primates (due to their close relation to humans) it has also promoted an undercurrent of presumptions about animal behavior.

Evolution through natural selection implies that a being (human or animal) acts the way it does so as best to protect its reproductive future. Any behavior that is exhibited throughout generations is there because genes so dictated. Does this mean that all behavior is genetic? To assume this would in a sense deny all forms of intentionality and culture among nonhumans. Of course this can only be assumed if one conveniently leaves humans out of the equation, which is most likely to happen anyway, even if it makes little sense. This notion would imply that culture, tool preparation, altruistic behavior, and even sign language is only observed in nonhuman primates because they are acting on behalf of their genetic codes, like unconscious machines following a program. It's not too difficult to see how closely this image mimics the soulless automaton of the Cartesian era and promotes separation of human from nonhuman primates.

On the other side of the scale is anthropomorphism (assigning human characteristics and/or thought to a nonhuman). The proscription of granting other species undeserved inclusion into the sphere of humanity leaves researchers in a tough space, especially when the animal behavior under question is entirely mental or is so subtle as to be difficult to measure. The intentionality of a nonverbal being will always remain something of a mystery. As Barbara Noske writes, "No scientist can ever totally transcend his anthropocentrism, in that he cannot leap over his own humanity and the typically human perspective. In that sense our fellow apes remain unknowable."[149]

What's the solution, then? Perhaps, as Noske suggests, researchers should learn from the apes not by teaching them our language and imposing our cultural norms on them, but by immersing themselves in the ape culture and becoming one of them.[150] Living in their communities, eating their foods, learning their communications, and practicing their behaviors might really be the best way to understand nonhuman primates. It would be the truest, purest form of observation and certainly more direct than teaching primates human behaviors in order to learn about primate behaviors.

When it comes to discussions of primate culture, intentionality, and intelligence, much of what has been studied, proposed, argued, and conjectured strives to reveal something quite mysterious: what animals are thinking (and how they are thinking it). Nevertheless, there does not yet exist any objective method of determining exactly what animals are thinking at any given moment. For example, chimpanzees and capuchins have been observed in situations where it appears they are taking part in an organized hunt, although opinions vary on whether the individuals intentionally cooperate, with a shared desire for hunting success, or if perhaps the behavior is something more random or coincidental. The hunt's success does increase with the size of the hunting group, which may suggest that perhaps this is evolutionarily intentional. If primates are indeed cooperating, though, it would mean much in terms of evolution and anthropology, because it's widely believed that mankind has been so successful in mastering his environment partially due to early humans' ability to cooperate among themselves in pursuit of a larger goal.

It's important to note that, in cooperation research at the Great Ape Research Institute in Japan, chimpanzees would not cooperate together to reach a shared food goal, but they would work with a human being to reach the same goal,[151] and they would help human beings reach a goal, as well, even if the reward were not something desired by the chimpanzee. But to blindly assume cooperation between individuals of another species is a huge assumption, because it implies an impossible reading of intentions and comparisons to human thought processes. Since these primates are not humans, one cannot make decisions about mental status based merely on comparing resulting behaviors with human behaviors.

It would seem that a simple solution to unlocking the mysteries of nonhuman primates' minds would be solved with a shared language whereby they could communicate to humans their wants, desires, and even thoughts. This is where the much-debated sign-language studies of the mid-1900s came into play. These began in the late 1940s with Viki, a chimpanzee who was trained over three years, through much labor and physical molding of her lips, by Keith and Katherine Hayes to vocalize four words (*momma, poppa, up,* and *cup*). It is generally agreed that Viki spoke the words not because she understood their meanings but because she learned that if she spoke the words she would be rewarded with food. Human language did not come naturally to Viki, nor did she seem able to speak with any sort of ease or comprehension.[152]

The experience with Viki, and other similar early studies in which chimpanzees were raised as humans to see if they would pick up human language naturally, simply cemented the concept that apes cannot use a spoken language due to the natural formation of their larynxes. Specifically, nonhuman primates are lacking a bend in the vocal tract that is required to make the sounds of a spoken human language.[153] Humans are capable of uttering and comprehending about a hundred different phonemes (the sounds that comprise language), even though most languages use only half that amount. Chimpanzees, it has been found, are only able to create twelve distinct phonemes.[154]

As an alternative method of communication, researchers thought perhaps American Sign Language (ASL) would be a perfect fit for primates, due to their high intelligence, social natures, and dexterous hands. In 1964 William Lemmon, a primatologist at the Institute for Primate Studies in Norman, Oklahoma, purchased an infant chimpanzee named Lucy from the owners of an exotic animal show. Maurice Temerlin, a clinical psychologist and psychotherapist, was chosen (along with his wife, Jane) to raise young Lucy as a human child in an effort to see just how much of its innate chimpanzee qualities a youngster could lose in such a situation. Working with the first chimpanzee to be reared by humans past sexual maturity, Temerlin was especially interested in Lucy's sexual development and the degree to which it would be influenced by these circumstances.

Lucy had typical human childhood experiences, such as wearing clothes, receiving immunizations, learning to eat with utensils, and surviving as a member of a family unit, and she excelled at them all, perhaps even better than a human child would have. She was adept at preparing drinks for herself (both alcoholic and virgin) and using tools such as screwdrivers to disassemble objects. As she got older, Lucy's behavior continued to mimic that of a human, albeit a human without inhibitions, as she was fond of masturbating and staring at human male centerfolds in adult magazines that were supplied to her as part of the investigation. When she was later introduced to a male chimpanzee, she was understandably frightened, never before having seen another member of her own species.[155]

Graduate students of the Institute for Primate Studies, including Roger Fouts, who would later work with another signing chimp named Washoe, were hired to teach Lucy sign language, and she eventually learned to incorporate more than a hundred signs into her vocabulary. As other primates did after her, she would create combinations of signs to signify things for which she didn't know the proper words, such as signing *"cry fruit"* to express *onion*.[156]

As Lucy grew larger and stronger, special considerations had to be made for what Dr. Temerlin considered to be his "daughter." A special room was built for her, reinforced with concrete and steel. Although most parents wouldn't allow it for a human child, Temerlin permitted Lucy to drink alcohol, calling her an "ideal drinking companion...[who] never gets obnoxious, even when smashed to the brink of unconsciousness."[157] Temerlin tested Lucy's comfort zones by engaging in various forms of sexual activity in front of her. Many of Temerlin's methods are considered outdated and inappropriate by modern standards, and they would likely have seemed questionable to most conservative people at the time.

After ten years of habituating Lucy into every aspect of their lives, the Temerlins grew tired of the experiment, and it was terminated. Although Lucy was now a media darling (while interviewed for a New York Times article in 1974, she invited the ASL-signing interviewer, Boyce Rensberger, up into a tree[158]), she was sent to live at Niokolo Koba National Park in Senegal, something that the Temerlins felt was in her best interest (despite the fact that their "daughter" had actually been born into captivity in the state of Florida). Although one of her teachers stayed with her at her new home, Lucy grew depressed at the loss of her normal life and family, and she became seriously ill. She refused to drink, climb, and eat like the other chimps in her assigned social group in Africa, asking in sign language for help when she grew frustrated at not having enough food.[159]

As her human helper tried to draw away for fear of retarding any potential progress Lucy could make in her new life as a wild chimpanzee, Lucy learned to manipulate her emotions by signing every time she was hurt, pulling out her fur in desperation, and growing emaciated from starving herself. After this low point, Lucy did eventually learn to accept her new life. She started eating leaves and slowly growing more confident in living outdoors. After ten years of living as a wild chimpanzee in Senegal, Lucy was killed and her skeleton found near the compound, with the hands and feet missing, likely from poaching activity.[160] Lucy died at age 22, having lived her short life first as half human, then half chimpanzee.

The first successful ape sign-language study involved Washoe, a chimpanzee who was also raised by humans. The project was started in 1966 at the University of Nevada by Allen and Beatrix Gardner and then spearheaded by a young graduate student named Roger Fouts (who later also worked with Lucy), with the goal of designing Washoe's curriculum in response to the previous failures at teaching primates to speak a human language. The Gardners suspected (correctly) that nonhuman primates were physically unable to produce speech and, considering their physical and manual dexterity, might be more successful communicating with signs.

Although Washoe was raised as a human child, no spoken language was used in her presence; only ASL was used to communicate. During the five years that she lived with the Gardners and Fouts, Washoe's language ability progressed as would a human child's language skills. She not only used language to gain access to food, but also in more abstract ways that highlighted her desires and opinions, such as to ask for playtime or describe something for which she didn't know the proper sign. She learned more than 130 signs, and her researchers estimated that she understood three times that number.[161] Video footage of Washoe playing by herself revealed private signing (Washoe signing to herself, the equivalent of a human talking to himself), animation (pretending that an inanimate object is alive), and substitution (giving an object a new identity). The discovery of Washoe engaging in private signing and labeling (such as signing "cat" when she saw a photo of a cat[162]) disproved centuries of philosophical decrees that only humans are capable of thought.

At five years old, Washoe was forced to leave her human family when the study supporting her sign language lost funding. The project was moved to the Institute for Primate Research at the University of Oklahoma under the direction of controversial director Dr. William Lemmon. It was here that Washoe first saw other chimpanzees. She described them as *"black cats"* and *"black bugs,"*[163] revealing just to which species Washoe felt she belonged. Over time, however, Washoe grew increasingly comfortable with the other chimpanzees, even reacting with true altruism and compassion when she rescued a fellow chimpanzee that was drowning in the moat surrounding their enclosure.[164]

At this point in the study, Washoe's closest human confidante, Roger Fouts, began to doubt the efficacy of the old-school, dominance-heavy scientific community and its relationship to its animal subjects. This nagging feeling would not go away and eventually would cause Fouts to turn his back on animal research in general, but not before moving the Project Washoe chimpanzees to Central Washington University, where he was able to demand a new level of respect and freedoms for the chimpanzees he had grown to love and value in their own right.[165]

Later in life, Washoe was found to have taught sign language to her adopted son, Loulis, via modeling with his hands and signing on his body. (It's important to note that Washoe's caretakers were careful not to sign in the presence of Loulis, so they could see if he would learn sign language from Washoe. He started signing within eight days of being "adopted" by her.[166]) Chimpanzees in the wild rely heavily on teaching and modeling to pass down cultural and survival skills to successive generations, so it is not surprising that Washoe taught her adopted infant the language she herself grew up using to communicate, but this also proved that not only those chimpanzees raised in human homes could successfully learn a human language.

Skeptics of Project Washoe claimed that the findings could not be presented as true of general chimpanzee intelligence and ability, but only proved something about cross-fostered chimpanzees. Luckily, little Loulis (who was not cross-fostered) learned sign language the same way he learned other skills from his mother, such as grooming and nest-making, and the scientific community soon had to admit that these chimps' language abilities were indicative of their entire species, and not just a rare coincidence.

Washoe and Loulis were eventually joined by other chimpanzees whom the Gardners had taught sign language— Moja, Dar, and Tatu—allowing the researchers to examine how the chimpanzees, as a group, used sign language in their daily lives. It was discovered that Loulis had a tendency to sign to his playmate Dar on certain topics and sign with his mother on other topics. When play between Loulis and Dar got too rough, Loulis would scream for his mother and then point to Dar and sign, *"Good good me,"* as if to incriminate Dar as the guilty party of the fight.[167] Both applications—the topic specificity and blame games—illustrate that Loulis had a concept of the mental states of others, something he tried to use to his own advantage.

With Project Washoe, it was discovered that the majority of chimpanzee signing fell into one of three categories: play, social interaction, and reassurance (which is generally true of other chimpanzee communication in captivity, as well). Surprisingly, food was not one of the main topics of discussion, so the chimpanzees were not simply mimicking signs in order to gain access to meals. The chimpanzees' signing revealed how far back their memories could stretch. For example, Tatu asked for the annual Christmas tree during an early November snowstorm,[168] and Washoe greeted her old human family, whom she hadn't seen for eleven years, by signing their names.[169]

In 1971, a more conventional scientific study was embarked upon by Duane Rumbaugh at the Yerkes Regional Primate Research Center in Atlanta, Georgia, with a chimpanzee named Lana. Out of concern that the more casual, family-style primate language studies would not be accepted by the scientific community, he instead relied on empirical methods in his experiment and tried to minimize any subjectivity.

Lana was taught to communicate via a computerized keyboard. The benefit of the computer was that it removed all subjectivity from the research, since it would only reply to Lana's requests if they were formed correctly. Any type of body language and unspoken inference (potential influences that had peppered the critiques of previous primate language studies) were meaningless to the computer.

Soon Lana was not only able to use the symbols correctly to ask for things she desired but also to describe things for which she did not know signs and to argue with her trainers when she felt they were teasing her.[170] While it's not surprising that Lana learned to use this language to her own benefit, it is interesting to note that she also learned to read output from the machine and to complete puzzles in which key words were missing from a sentence.[171]

Begun in 1973, Project Nim was a four-year attempt to teach sign language to an infant chimpanzee called Nim

Chimpsky (cleverly named after Noam Chomsky, a famed linguist who had proclaimed language to be the result of innately human qualities). Psychologist Herbert S. Terrace undertook the experiment in the hopes that it would help define the boundaries of humanity and its culture by exploring the development of human language in another species.

Nim was raised as if he were a human child, surrounded by a human family that was constantly supervising and recording his actions and under the tutelage of sixty different teachers. It was hoped that his close ties with the humans around him would inspire him to learn sign language as a way of communicating with and appeasing them. Unfortunately the project was plagued with instability—with many changes in personnel and locations where Nim was taught, as well as lack of funding—problems that surely did not help to calm the inquisitive and observant young chimp and most likely impeded his learning potential.

In the first four years of the project, Nim learned 125 signs, starting at four months of age.[172] The speed and accuracy of his learning depended greatly on the tenure of the teachers who were working with him. The longer they had been there to develop a relationship with Nim and the more patient and talented they were, the faster he learned novel signs.

Like other primates in language studies, Nim would combine signs to describe objects whose proper signs had not yet been taught to him. New signs were learned either by having his hands physically molded by a teacher or by his imitation of a teacher's signing. Because his teachers constantly communicated with and around Nim via sign language, it was easy to show that Nim was not using signs only in anticipation of a food reward. For instance, Nim used signs to ask for a specific color of crayon he wanted for an art project, to describe certain teachers by name, and even to identify himself in mirrors. When one of his teachers caught him about to drink some poisonous cleaning fluid that had accidentally been left out, her frantic signing of *"No stop don't eat"* halted Nim in his tracks and surely prevented a disaster.[173] Nim's sense of self and awareness of others appeared to be very well developed and documented.

Although Project Nim was later criticized for using outdated research methods and animal handling, including corporal punishment in times of extreme defiance, some interesting moments occurred that shed light on the development of young Nim. For example, his caretakers apparently felt that he was rather self-centered, acting as if everything in his environment existed solely for his use and enjoyment. Once his caretakers taught him the concept of taking turns and sharing, Nim began using the sign for *me* frequently; however, it wasn't until he had a tantrum one day after having to share yogurt with a teacher, when he angrily began using the new sign to chastise the teacher who (he felt) wasn't sharing adequately, that he added *you* to his vocabulary. From that point on, his signing reflected his adjusted view of the world, as one where actions and experiences could be shared between beings and where he was no longer the center of his universe.[174] This is a developmental step clearly delineated in human children, as well.

Other instances revealed Nim trying to teach sign language to human children he met outside, and he would also play tricks on his caretakers (such as hiding important objects they wanted) that revealed his clear sense of awareness of the points of view of others. He enjoyed cleaning and household chores so much that he would throw a tantrum when he wasn't allowed to participate.[175] During potty-training episodes, Nim was so accustomed to wearing pants that when they were removed to facilitate his use of the toilet, he covered himself up as if embarrassed by his nudity.[176]

Despite his seemingly human characteristics, Nim wasn't afraid to reveal his chimp-ness (for lack of a better word), whether it was by testing the will of newcomers with physical displays or screaming in defiance when he was displeased, which led to physical attacks on unfamiliar people. As revealed in the study, "Undoubtedly Nim saw a teacher who could not control him as an opportunity to assert his own meager dominance...."[177] "Most of the time, however, Nim expressed his feelings and intentions directly...This was true whether Nim was expressing affection, curiosity, fear, aggression, wonder, or determination."[178]

An interesting side effect of his sign language usage was that a few times when Nim was acting aggressive, he would approach the object of his aggression, generally a teacher, and sign the words *"bite"* or *"angry"* while looking as if he were about to attack the person physically.[179] This could be viewed as the language equivalent of a threat, something that is very pervasive in wild chimpanzees, albeit typically in the form of puffing out of body hair, physical displaying, and swaggering from side to side. Certainly, getting in the faces of his teachers, grimacing, and signing about how angry he was would instill fear in the humans around him. But Nim's use of *"bite"* and *"angry"* can also be considered as tools, as if his ability to express himself stopped him from relying on pure physical reactions—almost as if some of his chimp-ness had been suppressed and redirected into language by the human culture in which he was submerged.

One of the prevailing questions that Herbert Terrace aimed to answer with Project Nim was whether or not a primate was able to consistently and accurately create sentences with sign language. Although preliminary evidence seemed partially to support the hypothesis that Nim was understanding and creating proper sentences, this was never able to be examined in depth, and Terrace wrote that at times Nim's signing was more imitative and used solely to obtain positive feedback from his teachers. [180]

Project Nim ran out of funding in 1977, and the project was disbanded. The star of the show, Nim, had to be returned to the Institute for Primate Studies in Oklahoma, where he joined Washoe, who was also living there after being raised as a human and taught sign language.

At this point, successful experimental language studies, both with and without the usual scientific methods, had begun drawing attention, and that, plus federal funding, helped promote further investigations. After the relative failure of Project Nim, Herbert Terrace became skeptical of other ape language studies and was frequently asked to review video tapes and search for signs of prodding and imitation that were similar to those exhibited by Nim and his teachers. Terrace admitted that Nim was not taught true American Sign Language but a pidgin version; thus, even if Nim had learned it word for word, he would never have been considered the master of a previously recognized language.[181] Additionally, Terrace was very often skeptical of any investigation into animal language use that seemed to have positive results, although he was never able to explain certain events occurring in the chimpanzee language studies, such as the spontaneous signing of Washoe when alone with her dolls, or the signing of her son, Loulis, who used fifty-seven distinct signs, despite humans near him using only seven special signs.[182]

The year 1972 saw the first time that a gorilla was taught a human language. Francine "Penny" Patterson started teaching an infant lowland gorilla named Koko sign language as part of a research program through the psychology department at Stanford University. It took only a week for Koko to begin using signs to request food and drink, and as of this writing she has learned more than a thousand signs.[183]

As with other apes that have been taught to communicate using sign language, Koko's intelligence has been examined closely. She not only understands spoken English, as well as American Sign Language (she actually understands many more words than she is able to sign on her own), but she is learning to distinguish letters of the alphabet and has been tested as having an IQ of approximately 80 (which is considered low-average for humans.)[184] She recognizes herself in a mirror, has an active imagination, creates artwork and can communicate in the abstract about events that have happened in the past, using words to indicate a sense of time.[185]

Apparently quite empathetic, Koko expressed emotion when she saw photos of other gorillas or other animals that appeared to be suffering or in pain, and she bonded with kittens that she was allowed to keep as pets and deeply grieved their deaths. It's important to note that she wasn't merely grieving the absence of her beloved kittens; she actually understood the concept of death. When asked by her teacher where gorillas went when they die, Koko signed *"comfortable hole bye,"* and when asked how they feel when they die, she signed *"sleep."*[186] Koko had never seen a burial before, but this is consistent with the instinct of gorillas and other primates to cover the deceased in either earth or vegetation in the wild.

As Koko gained worldwide attention and admiration for her bright and gentle demeanor, Dr. Patterson established The Gorilla Foundation in an effort to guarantee funding for the duration of the research. Eventually two more gorillas named Michael and Ndume joined Koko at her home in California. Both additional gorillas learned sign language later in life and relied on it to communicate between themselves and to their human caretakers. The gorillas have become so proficient that they have been known to sign slowly to humans who don't understand sign language as well as they, also using modulation or the exaggeration of words to prove an especially compelling point, in much the same way that humans will slow their speech and overemphasize words to non-native speakers of their language.[187]

Among other items of interest, Patterson's work has shown that gorillas most enjoy eating; trees also make them happy, and work makes them angry.[188] Her gorilla subjects appear to harbor amazing creativity and keen observational skills regarding what's going on around them. For instance, Koko described the events of an earthquake as *"Darn darn floor bad bite. Trouble trouble."*[189] When asked how she had slept last night (a colloquial turn of phrase), instead of replying with a descriptive term about the quality of her sleep, as had been anticipated by the human questioner, Koko replied factually and literally with the physical description *"floor blanket."*[190] During a different exchange, Koko revealed her understanding of words in both concrete and abstract terms when she answered the question *"What is hard?"* with the signs for both *"rock"* and *"work."*[191] When someone jokingly refers to Koko as a goofball or some other such term, she corrects them and signs "no, gorilla."[192]

One of the most fascinating instances of ape humor was described in the following exchange between Koko and Barbara Hiller, one of her assistant caretakers.

> Koko was nesting with a number of
> white towels and signed *"that red,"*
> indicating one of the towels. Barbara
> corrected Koko, telling her that it was
> white. Koko repeated her statement
> with additional emphasis, *"that red."*
> Again Barbara stated that the towel
> was white. After several more
> exchanges, Koko picked up a piece
> of red lint, held it out to Barbara and,
> grinning, signed *"that red."*[193] [italics
> mine]

Ever the humorist, when Koko played jokes on humans, she would chuckle, sometimes even in anticipation of the actual events.

Koko and Michael have engaged in wordplay, using signs that were almost homonyms for the signs they actually should have used. "Knock" was used for "obnoxious," "tickle" for "ticket," "lip stink" for "lipstick" and "berry bottom" for "belly button,"[194] much in the way that young children will sometimes mispronounce or mistake a word for a similar-sounding choice. A study of Koko's language skills revealed that of the 876 signs she used during the first ten years of the study, six percent (or 54 signs) were invented by her. Two percent were compound signs she created of known signs, and one percent were considered "natural gorilla gestures."[195] Patterson says that "conversations with gorillas resemble those with young children and in many cases need interpretation based on context and past use of signs in question."[196]

Previous testing revealed that Koko could correctly identify and classify words and items, such as differentiating between light and dark colors or even describing some colors as warm and others as sad. Ninety percent of her matches were considered correct by the project's standards (by comparison, seven year-old humans taking the same study answered correctly just 82 percent of the time).[197] She continues to show self-awareness via self-recognition in mirror tests, converses about the differences between herself and others, and gets embarrassed when others observe her signing to her dolls in play.

The skills of Koko and Michael have not been considered exceptional or coincidental, as each came from different upbringings, and other gorillas in zoos have exhibited similar language aptitudes, albeit in naturally occurring gorilla gestures. Rather, it is believed that all gorillas have the aptitude to converse in American Sign Language if living in an environment that supports its development.

Chantek, an orangutan who was born in 1977 at the Yerkes Regional Primate Research Center in Atlanta, Georgia, was part of an American Sign Language study conducted by Dr. H. Lyn White Miles at the University of Tennessee at Chattanooga.[198] The research started when Chantek was just nine months old and proved that orangutans, in addition to chimpanzees, bonobos, and gorillas, were capable of learning a human language. Yet Miles was certainly not the first to recognize that orangutans may possess increased intelligence akin to that of humans. Lord James Burnett Monboddo, a linguist and anthropologist of the late 1700s, declared in his book *Of the Origin and Progress of Language*, "I still maintain, that his [the orangutan] being possessed of the capacity of acquiring it [language], by having both the human intelligence and the organs of pronunciation, joined to the dispositions and affections of his mind, mild, gentle, and humane, is sufficient to denominate him a man."[199] It wasn't until Miles' experiment that Lord Monboddo's theories could be proven.

Chantek ended up learning approximately 150 signs that were both unique and self-instigated.[200] He used the signs to gain things he wanted and to control the conditions around him, and when he did not know the particular sign for an object, he would combine known signs in a way to best describe the object he was discussing. Much like the other apes in sign language studies, and like humans learning language, Chantek would use specific words to describe a large range of related items: For example, *dog* could mean a canine or something with four legs or a picture of a dog or a noise approximating that which a dog would make.

Chantek also showed signs of utilizing displacement, discussing things which were not in his immediate physical world (something which is held as a marker of intelligence, as it requires symbolic and abstract thought as well as considerable memory).[201]

He developed a value system and could label what he considered to be good or bad, illustrating his ability to be enculturated into the norms of human society and what is and is not allowed by human individuals living among others. Chantek also learned deception and would often use signs to trick those around him into thinking something or acting in a certain way. He was able to think empathetically and consider the points of view of others, so as to best control his environment. Additionally, he would use animism and pretend that inanimate objects were alive. [202]

Unintentionally, Chantek also learned to understand spoken English and would respond to spoken inquires with sign language answers.[203] When prompted, Chantek could slow down his signing and deliberately make his language clearer to others[204] (like gorillas Koko and Michael would do). He also began to use his feet, as well as objects nearby, to sign. When he was eager to learn the correct sign for something, he would offer up his hands to the researchers as a plea for help, so that they might shape his hands into the appropriate positions.[205]

After nine years of living at the research site at the University of Tennessee, Chantek outgrew the facilities and was moved back to the Yerkes Center where he was born, where he stayed until 1997 when he was placed at Zoo Atlanta.[206] Although one assumes that Miles's work with Chantek was begun with an open mind and lack of preconceptions regarding the personhood of nonhuman primates, she admits that "like my colleagues doing similar research, I have found myself unconsciously experiencing them [the orangutans] as persons."[207] Miles believes that evidence of Chantek's language usage is sufficient as proof of rational thought,[208] and thus he meets the criteria for Descartes's definition of personhood.

A less rigid language study was performed in the late 1970s. Gary Shapiro was a young student who had been schooled in the methods and experiments of Allen and Beatrix Gardner and Roger Fouts during their work with the chimpanzee Washoe. He had also had experience working with Aazk, an orangutan being studied at a zoo in Fresno, California. He was nominated for a position teaching sign language to orangutans at primatologist Biruté Galdikas' study site at Camp Leakey in Borneo.

Shapiro arrived at Camp Leakey in 1978 and eventually established an emotional bond with a young orangutan named Princess, a former pet who had successfully been reintroduced into the wild and who ended up learning thirty-seven signs in nineteen months.[209] It's important to note that, in this setup, Princess was able to leave Shapiro and go off into the jungle at any point in time; thus, the project lacked many controls and structures that were deemed important in previous language studies, and Princess may have been able to learn and use more signs under different circumstances if there had been fewer diversions.

A few other researchers tried various methods to compensate for nonhuman primates' inability to utter the sounds of human language. In the late 1970s, David Premack, a psychologist from the University of California at Santa Barbara, pioneered a study that involved Sarah, a chimpanzee, learning language based not on spoken sounds or signs but on plastic chips. These chips acted as symbols for words, and throughout the study Sarah was taught not only to use the chips correctly but also to arrange them in a grammatical fashion, considering things like word order and ways to differentiate how some items were similar or different from each other.[210]

In another study, a bonobo named Kanzi, who now resides at The Great Ape Trust in Iowa, learned to communicate using an electronic board filled with various symbols. Each symbol stands for a unique word or concept, and primatologist Sue Savage-Rumbaugh, who heads the study, has documented that Kanzi has learned to use 350 words to express himself and has been able to combine them into a sort of "proto grammar."[211]

He is very proficient at understanding spoken English and comprehends a vocabulary of approximately 3,000 spoken words.[212] Further, Kanzi's sister, Panbanisha, was captured on video using chalk to draw one of the symbols on the ground in order to communicate to her keepers that she wanted to go outside.[213]

Savage-Rumbaugh has explained that bonobo symbol acquisition was only successful after humans stopped trying to teach the bonobos and, instead, simply used language around them. "The driving force in language acquisition is to understand what others (that are important to you) are saying to you. Once you have that capacity, the ability to produce language comes rather naturally, and rather freely."[214]

More recently, Panbanisha's infant, Nyota, has proven to be even more advanced than Kanzi and Panbanisha were at the same age.[215] This is most likely due to Nyota's growing up since birth among both educated, signing, human caretakers and signing bonobo family members, allowing him to live in a truly enriched environment that supports language acquisition. Savage-Rumbaugh's research with Kanzi and other bonobos has since been applied to helping autistic human children communicate with lexigram boards.

Common to almost all the advanced primate language studies has been the use of the word *dirty*. Interestingly, all the apes responded well to this word, and even though there were distinctions in how the word was applied to their worlds, the primates of the language studies found *"dirty"* to be a very expressive and useful utterance. The chimpanzee Nim used it to alert his teachers that he had to go to the bathroom, but he also used it in jest to distract his teachers during a lesson he found especially boring.[216] The chimpanzee Washoe used *"dirty"* as a curse word, to describe other primates with whom she was angry or who she felt had slighted her.[217] The gorilla Koko signed *"dirty"* to express her displeasure when her toy was accidentally destroyed.[218]

Another commonality of the language studies was the invention of novel signs to describe situations or objects. For example, Nim was sometimes given lotion to apply to dry skin on his hands. Eventually he started rubbing his hands together as a fabricated sign, which became known as his way of asking for hand lotion.[219] Washoe signed *"water bird"* the first time she saw a swan.[220] This, and numerous similar examples by other primates in language studies, seems to indicate that some species of primates are capable of creative expansion of language to communicate novel ideas. This ability to describe objects requires a true understanding of the language, although at times nonhuman primates, like human children, can take language too literally: When someone gave the bonobo Kanzi the task *"Put some water on the carrot,"* Kanzi threw the carrot outside into the rain. Nobody could argue that he didn't understand the command![221]

Depending on the source and age of the data in question, ape language studies have universally determined that nonhuman primate language is equivalent to that of a human child ranging in age from two to four years. Sue Savage-Rumbaugh claims that nonhuman primate language acquisition in studies is even more difficult than human child language acquisition because the input mode (spoken English) often does not match the output mode (ASL or lexigram buttons).[222]

Many characteristics of maturation revealed by the ape language studies, such as the capacity for deception, animism, and overextension of a word, were similar to stages experienced by human children of a similar developmental level but might occur with less frequency. Both the bonobo Kanzi and chimpanzee Nim failed to show the gradual increase in sentence length that is common among human infants learning language.[223] Kanzi's spontaneous lexigram communications revealed what his researchers referred to as a "primitive syntax or grammar…a protogrammar,"[224] or an ape grammar relative to Kanzi's brain function. This protogrammar reflected partial adherence to English word order, analogous to what a human eighteen-month-old would produce. A similar protogrammar was found in use by Ai, the chimpanzee in the Japanese mathematics testing program.[225]

The rush to explore primate intelligence through language was short-lived, however, and lasted only until the early 1980s, when general opinion asserted that language-using primates were the equivalents of highly trained performance animals and did not truly understand human language. This attitude existed despite a pivotal 1980 experiment by Sue Savage-Rumbaugh (of the Kanzi experiments) showing that chimpanzees were able to classify objects into specific categories, which proved that they understood the different meanings and uses for various objects.[226]

It seems apparent that apes in language studies are neither just mimicking signs nor merely the products of excellent training. The order of demands, open-ended questions, and spontaneous signing illustrate that nonhuman primates can and do use language to describe their environment in much the same way humans do, with very specific and accurate word choices. By the time that Savage-Rumbaugh was proving the extent to which primates comprehend and use human language, however, many of the stars of the earlier primate language studies had moved on.

After Washoe bit off the finger of a visiting scientist, who then threatened legal action, the sign-language-trained chimpanzees residing at the Institute for Primate Studies were deemed too much of a liability, and the University of Oklahoma ended that decade-long relationship in 1979. By the early 1980s, the chimpanzees there were sold to the Laboratory for Experimental Medicine and Surgery in Primates (LEMSIP) in New York.[227] Media stories drew attention to the transfer, and the public grew outraged and uncomfortable at the thought of language-capable beings spending the rest of their lives in research facilities.

In the end, the two most famous chimps, Nim and another signing chimpanzee named Ally, were brought back to live in Oklahoma, and while Nim ended up at a ranch in Texas owned by an animal welfarist, Ally was later placed in a private medical research facility in New Mexico. The paperwork for Ally's transfer has since been misplaced. The facility claims that the chimpanzee was received without a name, and he had since been renamed.[228] Only the language studies with Koko and her family at the Gorilla Foundation and with Kanzi's family at The Great Ape Trust survived the skepticism of the 1980s and are producing data to this day.

The accuracy of primate language studies has been in question ever since Nim's project director concluded that what had previously been assumed to be language comprehension was more likely instructor prodding and clueing. Researchers have been accused of being insufficiently critical of language testing, using physical or verbal cues to coax proper behaviors from the primates being studied, and finding what they wanted to find—and perhaps not what was necessarily there. To a number of people it seems obvious that, in many instances, nonhuman primates have learned the meanings of specific words, signs, or symbols, but critics find it harder to swallow the concept that primates have shown the adoption of any sort of grammar. Nonhuman primates may understand the meanings of words or signs, but evidence thus far does not show that words are consistently combined in any particular order.[229]

It's been thought that the relatively quick dismissal of the early language studies was a subtle attempt to keep human beings in their assumed place of dominance over the less intelligent "other" animals. Primatologist Geza Teleki put it well when he said, "We humans commonly react with astonishment upon discovering that chimpanzees can do something we consider special to humankind. Any evidence of intelligence overlap provokes the greatest skepticism, as the uniqueness of that quality in us is our most cherished illusion."[230] Conveniently, proving that an utterance or response is language relies on vague and tenuous definitions that vary with the individual; thus, this nearly impossible task is not unlike proving that another human is a conscious being.[231]

In order to come to an educated conclusion about the proper use of language by nonhuman primates, it's important to understand the basics of human language. According to George Yule, there are six special properties of human language that make it uniquely un-learnable by other beings, considered in terms of communicative and informative signals (intentional vs.

unintentional communication): displacement (the language may describe places and times other than those currently being experienced in time and space), arbitrariness (there is no natural connection between a linguistic form and its meaning in a particular language), productivity (the language can be used to create almost infinite word combinations, novel sounds, and new meanings), cultural transmission (language is learned from the surrounding community and passed on generationally), discreteness (sounds used are meaningfully distinct), and duality (language is organized on both the physical level and on the level of meaning).[232] Although some people may claim that the language skills of nonhuman primates are inferior and that their abilities are not sufficient proof of intelligence, it seems compelling to note that only in the last century did linguists cease judging languages of indigenous human populations to be inferior to that of "civilized" Caucasians.

The temporal lobe of the left hemisphere is the language-making center in the brains of humans, chimpanzees, and bonobos. Studies of ape gesturing have proved brain lateralization similar to that of humans, leading scientists to consider that language-making capabilities may also be similar, despite the fact that human brains are three times as large as the average brains of other primates.[233] Roger Fouts, who worked with the signing chimpanzee Washoe for many years, came to understand that great ape education must be spontaneous and that it is unpredictable, for the apes can sense when it is forced and, almost as if to prove their independence and freedom from any human-imposed restrictions, will not learn on schedule.[234]

In his essay "Rational Animals," Donald Davidson expresses difficulty imagining that animals could have much thought at all without language.[235] If we assume that other primates have thought, do they also have language? Critics of primate language studies, such as cognitive scientist Steven Pinker and esteemed linguist Noam Chomsky, believe that human brains are host to a physical element unique to our species, despite the fact that such a "language acquisition device" (LAD) not only has never been located in a human brain and that most scientists believe instead that language is a product of various parts of the brain working in tandem. Additionally, if there were a LAD in human brains, evolutionary theory would dictate that there be a similar element in the brains of other primates, albeit perhaps in a more primitive form.[236] Primatologist Tetsuro Matsuzawa lies somewhere in between the critics and the primate language proponents, explaining, "I do not say chimpanzees have language, they have language-like skills."[237]

Narrowing the definition of language to the point of requiring that language use be defined as limited solely to those beings that can produce it vocally would be to disregard the vast language capabilities of mute human beings, for example. Perhaps the distinction should rest with language comprehension, which page upon page of research over the years has proved to be an ability exhibited by nonhuman primates.

What the language studies, and others pertaining to any aspect of development or life of a nonhuman primate in captivity, do show is that results will always be atypical in that they are in some way specific to the given research subject. There are many factors involved in research which, added together, conclude with a subject leading a life quite far from natural. For example, as part of Project Nim, its central study subject was taught to complete routines that were important culturally to humans but were completely worthless to a nonhuman, such as hanging his hat and coat on a hook upon entering his classroom. Nim had to concentrate on actions that were evolutionarily worthless to him and were presented to him only as vehicles to introduce new signs and also to fully enculturate him as a young human, and his resulting outbursts and lack of interest seem to illustrate that he must have experienced some frustration at being forced to complete such unnatural tasks. Frustration apparently affected his learning curve and his patience with sign language; thus, the language capabilities of a nonhuman primate brought up in a laboratory or private home might be greater than or less than a member of the same species leading a natural life in the wild.

Although some researchers credit the language research environments with being more enriched and supportive towards the growth of impressionable young primate minds, as compared with typical life in the wild, Craig Stanford proclaims, "researchers must make an a priori assumption that they are studying a socially and psychologically stunted animal."[238] Of course, there's no way to test the language capabilities of the same being in the wild, simply because that would require them to leave the environment in question, and the results would inevitably be skewed one way or another. This is one of the great moral quandaries of using living research subjects: ironically, the more curiosity there is about animals, the more intensely they must be subjected to our scrutiny at the cost of their natural lives.

Even though the studies promote and celebrate the various levels of primate intelligence, language studies can cause controversy with animal welfarists. Experimental protocol often acclimates the subject to a comfortable life that is heavily influenced and altered by human cultures. The subjects can develop powerful emotional bonds with their teachers, especially in the case of studies that go on for decades. If a language study ends by the researcher's request or an account of a lack of funding, the study primates may be unable to acclimate to any other life, even the relatively more "natural" environment of a zoo or sanctuary. Certain primates are already so genetically similar to human beings that it's not difficult to imagine that their behavior would grow more human-like when living among humans, completing tasks that are normal to humans but are completely abnormal to, and useless for, primate life in the wild.

Habituating a wild animal to the life of a human is also understandably controversial because if, after decades and decades of primate language studies, it is determined that primates are fundamentally unable to foster any sense of human language (an unlikely assertion), many primate lives would have been spent in vain; yet, it's possible to view primate lives as spent in vain regardless of any discoveries about primate language. Clearly primates can communicate naturally and quite well among themselves in the wild, and they certainly do not need humans to pioneer research into their language capabilities (for their own benefit).

It would be ignorant to assume that only a human language counts as evidence of the language capabilities of nonhuman primates. The high level of intelligence and sociality of nonhuman primates, especially the great apes, makes them seem prone to developing a language that would evolve over the years; yet, apparently this hasn't happened nearly at the same rate as human beings have developed language. Reasons proposed for this include ideas such as nonhuman primates' lacking a theory of the mind (the ability to recognize that others have awareness and thoughts), or lacking a desire to communicate because there is no evolutionary benefit to it. Primates, like many other species, have developed alternate systems to read the behavior of others so as to work together on survival.

Many researchers have documented evidence of innate primate language. Bonobos have distinct hooting calls that are very different from their close relatives, the chimpanzees. These "echo" calls do not overlap and appear to be a sort of back-and-forth exchange in relation to some event, much like the language of human beings. Primatologist Frans de Waal postulates that the more peaceful nature of bonobos results in their communications being more vocal; thus they exhibit fewer physical displays (loud and violent stamping, waving and rushing emotional outbursts) than do chimpanzees.[239]

For twenty years, Dr. Klaus Zuberbühler of the University of St. Andrews in Scotland has been studying the calls of a population of Diana monkeys living on the Ivory Coast. He has found that the monkeys make specific calls for specific threats, be it a ground predator, such as a leopard, or a flying predatory bird.[240] Research by Robert Seyfarth and Dorothy Cheney of the University of Pennsylvania has proved the existence of a similar innate language system in vervet monkey and baboon communities. Individual vervets will even play a sort of practical joke on others of their community by making false land-predator calls. This compels other group members to hide in the trees, often leaving the caller free access to something he desires, be it food or a prime location at a drinking hole.[241]

The innate languages of monkeys and apes do not appear to be as malleable or creative as human language. Some species such as vervets have a more stringent vocabulary, being unable to combine sounds to make new words with distinct meanings, and do not appear to have separate calls to distinguish details of, say, the proximity of a predator. Conversely, Campbell's monkeys of the Ivory Coast have been observed to add suffixes to their calls in an attempt to differentiate calls about predators they observe themselves from those about predators they learn of via the alarm calls of a neighboring group of Diana monkeys.[242] This may be a way of separating proven fact from gossip.

Both Campbell's monkeys and putty-nosed monkeys will also combine sounds to make words with very distinct meanings.[243] The order of calls suggests a type of grammar, as specific meanings are dependant on the order of calls emitted by the vocalizing primate. For example, when Seyfarth and Cheney played a recording of baboon calls that had been edited so as to make it appear that an infant was threatening an adult (a complete reversal of the normal hierarchical order of their society), the baboons listening to the call looked at the speaker in confusion. The language they heard was not making sense, and they understood the absurdity of the situation. However, understanding a sequential order of words does not lead them to speak or vocalize accordingly in response. "The ability to think in sentences does not lead them to speak in sentences," the researchers explained.[244]

It seems that each time something new is discovered about the abilities and intelligence of nonhuman primates, a plethora of questions arises as a result. Does tool use imply culture? Can empathy exist in animals surviving in the wild? Language may be the result of detailed mimicry...or is it a window into the thoughts and preferences of a previously unknowable being? To answer such questions, is it best to increase funding for research, or is this quest a selfishly human desire that can have no benefit to the research subjects?

Ethical treatment of nonhuman primates is discussed at length in chapter eight of this book, but even if the entire book dealt with the subject of ethics, it would likely be impossible to come to a conclusion universally accepted by the myriad industries with an interest in nonhuman primates and their welfare. Humanity has been confounded by nonhuman primates through history, and the men of the land have still not quite figured out the best way to handle the men of the forest.

3. The Pet and Entertainment Industries

The fact is that monkeys and apes make us nervous, and they sometimes make us laugh, because in looking at them we see ourselves. They share, to one degree or another, a shape we regard as the honored shape. Their faces and bodies mirror our faces and bodies. But what we fail to see—or what we see only imperfectly and with some anxiety—is that their minds mirror our minds, as well. The final irony of people laughing at a group of endangered great apes is that great apes are the only animals in the world sophisticated enough, sentient enough, to laugh back. [245]

- Dale Peterson

In the United States, the legal status of nonhuman primates in the pet and entertainment industries is muddy. Various laws attempt to regulate the use of nonhuman primates in the pet and entertainment trades, but local attitudes and legal priorities can vary so widely that what is illegal in one state may be perfectly permissible in a neighboring state. Although the Animal Welfare Act dictates that the United States Department of Agriculture (USDA) is responsible for regulating certain ways in which primates can be held in captivity, such as for sale or on exhibition, the USDA does not in any way oversee individuals who own primates as pets.

Even if federal regulations permit such a thing, state and local jurisdictions have the ability to regulate the permissibility of keeping primates as pets, and ordinances vary widely between states and regions. As of the writing of this book, 22 of 50 states have a complete ban on nonhuman primate pets, and four prohibit great apes as pets but permit smaller primates to be kept as pets. Eight states require permits to have a primate pet, and 16 have no regulation at all in regards to primate pets.[246] Of the states with bans, many have clauses to grandfather in pets that were already in the state at the time of the legislation's passing, and some also make exceptions for people who move into the state with a primate pet that was obtained legally in their previous state of residence. The Humane Society of the United States estimates there are currently 15,000 primates living in private homes in America. As most of these rulings were made from the 1990s through the year of this writing (2012), and primates can have life spans of up to fifty years, there is the potential for large numbers of grandfathered-in primate pets living in areas where keeping them is supposedly banned.[247]

The keeping of primates as pets and as labor has been going on for thousands of years in areas of the world where nonhuman primates live in nature and can be put to work. In such areas they have been used to help harvest fruit or find water. In the modern culture of the United States, however, the relationship between humans and their pet primates is quite different.

Some people seek out pet primates in order to find a perpetually needy replacement for a human child. The National Geographic program *My Child Is a Monkey* claims that most owners of primate pets are mothers who lament that their human children have grown up. A woman identified as "Audrey" is such an example. The owner of two capuchins, she explains, "I'm a monkey mom. That's my identity. I don't have an [other] identity, I'm just a monkey mom. They're not animals to me. They're like little hairy people."[248]

The desire to acquire a nonhuman primate as a pet can arise out of a true passion and interest in the animal, or it may be related to the prestige of living with an exotic status symbol. Some primate pet owners admit to getting a thrill as they live out their Tarzan fantasies, and others believe that keeping endangered species as pets helps protect those species for the future.

The "endangered species defense" is often used similarly to justify the breeding and selling of primates as pets. Of course, unlimited breeding will ensure that particular primate species don't become extinct, but the unstable quality of life that many pet primates lead may seem, to some people, even more depressing than a complete extinction of a species. A lack of regulations means that owners have great discretion in how they raise their pet primates. Pets may live in dank basement cages, with access to only the bare necessities to maintain life, or be forced to wear frilly baby clothes and diapers and eat what humans consider *haute cuisine*. It all depends on the desires, considerations, and pocketbooks of the owners, and very rarely are the wants or needs of the pets considered at all.

Although importing nonhuman primates into the United States became illegal in 1975 (see chapter six, Legal Progress, for more information on this subject), where there's money to be made, some people will find ways to circumvent the law. Instead of exporting monkeys and apes from places where they were captured in the wild, breeders then opened shop in the United States and bred the primates that had already made their way into the country, thus ensuring a continued supply for the market.[249] In the breeding business, baby monkeys or apes are removed from their mothers when they are days or even just hours old. The younger they are, the easier they are to handle, so the infants are more valuable to breeders.

In the wild, primates are genetically conditioned to grasp and hold onto their mothers continuously for years. They are denied this close relationship when breeders separate mother and infant. As a result young primate pets are often found clutching blankets and soft toys, hugging them close in an unconscious search for a surrogate mother. If this doesn't cause enough Freudian dilemmas in young primates, it's well known that females who have been removed so early from their mothers tend to reject their own infants in the future. This is one breeding industry method that leads to a quite predictable cycle of poor parenting through generations of nonhuman primates.[250]

Despite the real, long-term difficulties that await the

purchaser of an infant pet primate, breeders and their customers will commonly seek out infants, specifically because in the beginning months of life they are tiny, helpless, and quite simply adorable. Short-term satisfaction is often the reason behind risky behavior, and this practice is no different. Breeders are in the business to make money, and they make the most money when they sell infants that are days old and have just been removed from their mothers, regardless of a customer's ability to provide appropriate lifetime care to the new pet.

It has been suggested that if breeders were truly interested in helping primates as a species, they would run a sanctuary, donate money to the conservation of wild primates, or get involved in activism on behalf of primates' legal rights. Instead it seems that most breeders operate their businesses in ways that mislead the public about the difficulties of primate pet ownership. In a frank, factual article in *Southsider Magazine*, primate veterinarian Craig J. Blair discusses how one of the most prolific pet monkey breeders, Rainbow Primates, bragged that their employees spend up to two hours educating pet owners about their new monkeys. Blair says, "I've been in exotic pet practice for 11 years and received training at the Cincinnati Zoo. I can say with certainty that two hours is barely enough time to realize that you will never know everything you need to know about monkeys."[251] The pet primate industry makes its living off idealism that often ends, unavoidably, in a stark and upsetting reality.

Unfortunately, the qualities that make some primates so endearing to human beings—their intelligence, physical dexterity, and unique personalities—are precisely what threaten their well-being when they are sought out as pets or entertainers. Sadly, these clever beings can suffer tremendously when held in captivity, simply because they are more aware than other pets of their own desires and the limitations of their surroundings. For example, one orangutan that had been raised by humans but ended up at a sanctuary actually seemed to prefer living as a human. As she was forced to live the life of an orangutan (most likely for the first time ever), she would sweep up messes, sharpen the blades of human tools, chop wood, and even attempt to start fires as she had seen humans do. Once she tried to present her meal of rice as a gift to the head of the sanctuary. Although this seemingly altruistic behavior may have appeared to be designed solely to achieve brownie points, the staff recognized the shrewdness of the act, for the orangutan didn't like rice and was trying to give it away, not out of generosity or just to curry favor, but as a covert way to avoid eating something undesirable.[252]

The concerns about keeping primates as pets begin at the source: their provenance. Depending on local laws, pet primates may currently be purchased from pet stores or individual breeders. Those animal suppliers all obtained their primates through the existing United States exotic animal network, since due to 1975's Convention on International Trade in Endangered Species of Wild Flora and Fauna (CITES) it has been illegal to import primates into the country.[253] As it operates now, though, there is much left to be desired in the exotic animal trade. Because the same animals may be shuttled among zoos, disreputable for-profit sanctuaries, and individual owners, learning the travel path or destination of a particular primate can be nearly impossible. The system in place to track the provenance and destinations of exotic animals is weak and overworked, and accreditation is shoddy and often nonexistent.

Basic internet searches can lead to advertisements for primate pets for sale by established and newer traders. The primates may have been bred in the US, where infants are removed from breeder mothers as soon as possible so the mothers can regain their fertility quickly, or they may have come to the US illegally in one of the many transports that escape the prying eyes of the International Primate Protection League in airports in Taiwan and Borneo. (For more information on this subject, read chapter seven, Primate Protectors.) These future pets may have had companions that died in transport from shock and lack of sufficient food and water, and they may be carrying diseases from their native lands.

Surplus apes and monkeys that don't find immediate homes as pets may end up as breeders for the dealers, so long as they haven't yet been castrated or spayed. Thus they unwittingly perpetuate the cycle whence they came. There seems to be no shortage of pet primates for those interested in buying them. April Truitt created the Primate Rescue Center, her sanctuary in Kentucky, when she become involved in the animal auction industry and learned the truth about where her pet monkeys came from: "...that was before our monkeys started biting, and before they became aggressive, and before we realized that the secret of the whole monkey trade is that there are no adult pets. Then we started finding these animals in basements; after they had bitten someone, people would just chain them up and stick them in a birdcage or something. There were nothing but unhappy endings.... And that was the beginning of it. Once you start looking and a couple of people get your number, the animals are everywhere. They are just literally everywhere."[254]

A quick search will turn up ads that promise the equivalent of a hairy human infant, ready to be dressed in baby clothes and taken out in public, for a few thousand dollars. One of the details left out of such advertisements is that these infant primates, who were taken from their mothers prematurely, were never able to develop natural bonds as they would in the wild. This leaves lasting impressions as foundations for the behaviors they develop later in life. Dr. Shirley McGreal of the International Primate Protection League has seen what happens regularly when primate pets outgrow the human enculturation that has been forced upon them. She explains, "That any human, male or female, could imagine they could provide the quality of care for a baby monkey that a monkey mother does is delusional thinking. Yet many people do imagine that any monkey raised by a human is lucky."[255] The surrogate mothers of their infancy are just one example of the abnormal life of a primate pet.

Another thing the ads don't tell people about, of course, is the mental stimulation and physical activity necessary to raise a primate to be even halfway healthy mentally. Ron Winters, a chimp owner and animal trainer, explains, "...a chimp is smart.

He knows he's not supposed to be in a cage. They hate that because their mind is too intelligent. They're bored to death."[256]

Primates are adept at breaking locks to steal, eat, or destroy personal items. They can be loud and messy, requiring the constant cleanup of their living quarters. Some primates can wear diapers, but even then, living spaces soon smell bad, and the larger and older a primate gets, the more he or she will resist the human intrusion of diapering and even of cleaning. Socially conscious behavior does not exist to them, and there is often no question when they start to mature sexually. Behavior of teenaged primates is often overtly sexual and rarely inhibited. This is only the beginning of care for a healthy primate pet.

The costs and amount of care required if the animal develops an illness can skyrocket. In 2009, the lifetime cost of care for a small capuchin monkey was estimated to be at least $70,000,[257] and the upkeep for a chimpanzee cost at least $15,000 per year (in addition to the purchase price of around $50,000), all of which can be assumed to rise in subsequent years.[258] They are susceptible to catching human illnesses such as measles and mumps and, depending on their origins, may have contracted more serious illnesses in their past. A healthy pet primate should be fed a diet of 70 percent monkey chow (a balanced mixture of dry nuggets that nutritionally mimics the diet of a wild primate) and 30 percent fresh fruits and vegetables,[259] something many owners are either ignorant of or unable to afford, or that they do not desire to follow, as it easier to feed their primate pets the carbohydrate-, sugar-, and salt-heavy diet of a typical American human. Just as it does eventually for humans, such a diet will make any other primate ill.

Most potential owners of primate pets are not exposed to accurate information describing what it takes to "ape-proof" a home or how often and how quickly a primate can become aggressive. The *Primate Care Handbook* by the Simian Society of America (a group promoting responsible primate pet ownership) realistically describes a popular cycle in the pet trade: "There are perhaps a limited number of individuals suited for the task of long-term primate caretaking. Unfortunately, the story of 'see monkey-want monkey-buy monkey-tire of monkey' is all too familiar... and it is usually the monkey who ends up losing the most for our misjudgment in taking on such a responsibility."[260] Indeed, Heather Thomas, Education Coordinator of the Duke University Primate Center, goes so far as to say, "It is true that primates cannot be domesticated."[261]

Primates in the wild have species-specific schedules for the amount of time they prefer to complete their natural behaviors, such as food gathering, resting, grooming, and playing. Although it is commendable that some primates living in captivity are given the chance to complete some of these natural behaviors, the inherent differences between their captive living spaces and the environments they would inhabit in the wild means that the behaviors are not completed the same way. For example, in the wild, foraging for a day's food may be completed slowly, while walking or climbing miles over the course of the day. When the same animal in captivity forages for even half the amount of time, without walking or climbing, it is likely to succumb to the obesity epidemic that plagues primates in captivity (especially the great apes). In the wild, grooming is a naturally occurring practice, but when captive, some individuals will groom to the point of baldness. This, like obesity, is most likely due to the absence of other natural factors in their captive lives, such as adequate exercise or sufficient mental stimulation.

The National Research Council's inquiry into the appropriate handling of primates in captivity, *The Psychological Well-Being of Nonhuman Primates*, delved into much detail involving the natural tendencies and preferences of primates. One of its recommendations was to allow primates living in captivity to control as much of their environments as possible. Since in the wild they complete "work" tasks and occupy their time in this way, it benefits them also to have tasks to complete in captivity. Often called "enrichment," these activities mimic foraging for food through play with puzzles and objects they can manipulate and use to relate to their surroundings. The keen observation skills of primates is also related to why they can be difficult to handle, as this treatise states: "When a routine is essentially innocuous to the animal, clear signals that identify an environmental change as part of a familiar routine can be beneficial.... Similarly, when procedures involve some level of discomfort to an animal, long anticipatory periods signaling the impending event can be a source of distress."[262]

The document goes on to explain that because primates are generally very social, they should be housed with others of their species or, if that's not possible, within visual range of others of their species. It states that most primates resist handling,[263] and although many species have been successfully trained to cooperate with medical procedures, such as moving to a particular cage or presenting an extremity for blood extraction, primates in captivity are often handled quite a bit more than is comfortable for them. In this unnatural environment, a frequently handled animal who naturally resists handling by humans is bound to experience tension and mental stress. Perhaps this is why the National Research Council admitted that "providing for an infant's physical needs is far easier than providing for its social needs."[264]

Although it's very rare for wild primates to attack humans, it's much more common for captive-bred primates to exhibit aggression against nearby humans. As primate expert Dale Peterson explains about chimpanzees (one of the most common captive primates in the United States), "Having placed chimpanzees in a highly abnormal confinement, in short, we discover that they resist confinement. They become dangerous, and we suddenly realize we must protect ourselves from the monsters we ourselves have created."[265] Pet owners can be reluctant to report injuries sustained or attacks by their primate pets, due to fear of prosecution, which could involve losing possession of the pets. This is why pet primates' teeth are often removed.[266]

Aggression during play, while harmless when exhibited by an infant, can turn bloody and violent when primates advance into teen and adult ages. Primatologist Richard Wrangham explains, "They are really a bridge between ourselves and other animals. They are, perhaps, caught between two worlds. They are capable of thinking quite advanced thoughts about what other individuals around them are doing, and yet they're not great at inhibiting some of their more powerful emotions."[267] Additionally, the intelligence of primates and their ability to intuit the emotions of their owners means that they can selectively destroy cherished objects to get revenge, gain attention, or simply release some energy.

Primate cognition expert Herbert S. Terrace seemed to question the very safety of his sign language project involving Nim, a chimpanzee raised within a human family, when he described Nim by saying, "At times he seemed so excited that I worried about the way he sank his teeth into the base of my neck. Even though I was sure that Nim was not trying to hurt me, I wondered if he knew the effect that his 'love nibbles' might have on my human skin."[268] Later, when describing how Nim would exploit his new teachers in order to proclaim his dominance, he wrote "It proved impossible to prepare new teachers fully for the quickness and the ferocity of his outbursts of aggression. Despite extensive briefing about what to do when Nim attacked and extensive opportunities to observe others struggling to control him, new teachers often felt quite helpless when their turn came. Nim was amazingly quick and agile. Having defended themselves successfully against one attack, his teachers were simply unprepared for the speed with which he would attack again, usually in a different manner."[269] This kind of turbulent excitability is not rare behavior in a pet primate, either.

Like any species, humans included, it can be difficult to know exactly what causes aggressive or antagonistic behavior, unless an individual is able to describe exactly what provoked the extreme reaction. One would think the primate sign language studies may have been able to help with this, as the subjects supposedly learned methods of communicating their desires and opinions; however, all Herbert Terrace was able to prove was that "Nim's aggression increased mainly because of the necessity of introducing more and more teachers into his life. The sheer frustration of losing a caretaker replaced by a new person was sufficient to evoke Nim's anger and aggression."[270]

Pet primates are known to complicate existing family dynamics, usually by seeking the approval or attention of one member and even ignoring or downright physically attacking another family member. It's easy to see how this can cause rifts among the humans living in such an environment. This human frustration is easily noticed and absorbed by the primates living in the strife, who then may have additional tantrums, contributing to a cycle of discontent with little respite.

At Camp Leakey in Borneo, primatologist Biruté Galdikas experienced similarly volatile behavior with Sugito, an orangutan that was orphaned after his mother's death and was then raised by Galdikas. Galdikas was married at the time, and her husband, Rod Brindamour, had been living with her at the research site. Despite knowing both Galdikas and Brindamour for equal periods of time, Sugito acted out against Brindamour through physical attacks on him and any other person or object he seemed interested in... except, of course, Galdikas, who was the closest to a mother figure that Sugito had.

Galdikas realized that from the point of view of a wild orangutan, it would be completely unnatural to have a dominant male around all the time, for in the mostly solitary orangutan society, a male does not stay with a female after mating. It's easy to believe that Sugito was experiencing jealousy and confusion at this abnormal occurrence and that his acting out was a reflection of what he considered to be the natural order of things.[271] Over time, Sugito's violence escalated to the point at which he was killing kittens and suspected of drowning other orangutans who had earned Galdikas's attention. The abnormal violence may be attributed to psychosis from observing the death of his mother and being raised in a culturally irrelevant way, but it also seems clearly tied to wanting his "mother" Galdikas's full attention.

It's a compelling theory that the search for a mother's undivided attention may be the cause behind the sometimes inexplicably violent outbursts of primates living as pets. Certainly it seems to be the case in other instances, as well. Consider the circumstances regarding Travis, a 14-year-old chimpanzee who had been purchased from a breeder for $50,000 and was raised by Sandy Herold in Stamford, Connecticut, as a pet and surrogate child. Travis was known around town for his human-like antics, such as eating and drinking with silver and glassware, riding around in Sandy's trucks with her, and using a computer.

On February 16, 2009, Sandy's good friend Charla Nash came over to help cajole Travis back into the house after he had escaped, and Travis viciously attacked Nash, seemingly for no reason. The bites and injuries to her body and face left her blind, seriously disfigured, and missing digits. It was considered a miracle that she survived the attack at all. The news outlets sensationalized both the violence of the attack and Travis's luxurious upbringing by emphasizing that he wore human clothes, drank wine from a glass, and slept in a human's bed. Herold claimed to be just as shocked as everyone else about the sudden attack and said that Travis had never before shown signs of doing anything like that.[272]

When that case is compared with that of Sugito, the orphaned orangutan adopted by Biruté Galdikas, it seems possible that Travis may have suffered from the same cultural frustrations. Although chimpanzee societies are much more social and complex than that of orangutans, male chimpanzees, like male orangutans, do not form special bonds with the children they sire. Although they may remain in the same social group as the mother who raises their infant, they do not bond with their kin, and the mother raises infants mostly alone (any help received usually comes from other females in the group). Travis was raised for much of his life as an only child by Sandy Herold and formed a tight bond with her. It's possible that he saw Nash, Herold's close friend, as a threat because she took away his mother's attention. Although he may not have attacked her in the many other instances when the three of them were together, at the time of the attack Travis was loose outside and perhaps sowing a bit of his wild oats. He was also suffering from Lyme disease, and although few cases result in psychotic episodes, Herold later admitted that she had given him some Xanax medication to calm him earlier in the day, and he may have been reacting to his illness or the medication. Travis was shot dead when he rushed at a responding police officer after attacking Nash,[273] but even if he were alive today, there would be no way to get answers as to why the attack happened. As illuminating as any potential answers in this case might be, they would still not be applicable to all instances of violence committed by pet primates.

Aggression and destruction are only part of the many problems associated with keeping nonhuman primates as pets. As with many other exotic animals, oftentimes the owners do not keep them in what many people consider to be appropriate living conditions. Primate sanctuaries are all too familiar with rescuing primates from filthy homes where they received substandard care and attention. Diets lacking the proper nutrients, improper temperature and humidity control, being kept away from sunlight and the natural world, and insufficient mental and physical exercise results in pets that are mentally and physically ill. Their health issues are often seen at first glance: they lose fur, can become emaciated or obese (sometimes resulting in uncontrolled diabetes) and may grow listless, with little interest in their surroundings. Males in captivity show a high rate of stress-related heart disease, which does not occur in the wild.[274] Nervous disorders such as tics, ulcers, and chronic diarrhea may develop. Due to the risk of injury and disease transmission, many veterinarians refuse to treat primate pets, and it can be difficult to find a veterinarian who has any expertise in treating primates.[275]

Macaques, members of a primate species that is popular in both the laboratory and pet industry, are natural carries of the Herpes B virus (Herpesvirus simiae), which can be fatal if contracted by humans. It has been estimated that approximately 80-90% of macaques carry the virus.[276] Macaques with Herpes B may not suffer any symptoms, but unlike other viruses that won't jump the species barrier, it can be very easy for humans to contract Herpes B from contact with an infected macaque. From a splash of bodily fluid to a bite, once humans are infected with Herpes B, they quickly become very ill, suffer neurologically, and may die of brain inflammation. Approximately 70% of human infections are fatal, and most infected humans suffer neurological damage.[277]

This information is intimidating, to be sure, but what's most frightening is that an uninformed pet owner or roadside zoo visitor may never know what risks are at hand when handling a macaque. This, coupled with any innate aggressive tendencies a macaque may harbor, can result in some upsetting and potentially very dangerous situations. Terri Parrott, a Florida veterinarian, spoke out about Ringo, a macaque who was seized from his owner after he bit a child (who later died). She says that anyone who owns a macaque as a pet faces a dire—albeit it largely unknown or misunderstood—threat. "If I hit Ringo, he'll attack somebody else, because that's just their pecking order. My children, who are seven, five and four, come in here, and if I yell at Ringo, he tries to attack my kids. That's just the way macaques are. What's going to happen," says Parrott, describing unsuspecting pet owners, "is that the parents are going to yell at the baby macaque, the macaque is going to bite the kid, and the kid is going to come down in eleven to twenty-one days with flu-like symptoms. Then they'll take him to the pediatrician, where he won't get treated correctly, and he'll die."[278]

Macaque-harbored Herpes B is not the only health threat that can be communicated between human and nonhuman primates; bite wounds can cause wound infections and extreme physical deformities, and Ebola virus, monkey pox, tuberculosis, hepatitis A, shigellosis, cholera, and simian immunodeficiency virus (SIV, a relative of HIV that can cause AIDS in humans) can be shared, as well.[279]

At a 1998 annual meeting of the Conference of State and Territorial Epidemiologists and National Association of State Public Health Veterinarians, exotic pets' effect on public safety was discussed. Communicable diseases common to pet primates, such as Herpes B, simian immunodeficiency virus, and Ebola, are not preventable with human immunizations; likewise, humans have passed Hepatitis A, shigellosis, and tuberculosis to their primate pets.[280] The meeting concluded with the decision that the group would support making private ownership and future breeding of nonhuman primates illegal, and that any pets already grandfathered in from previous legislation be prohibited from being in any public space and never be bred.[281]

National Geographic's 2009 program *My Child Is a Monkey* provided an in-depth view of people who are driven to keep primates as pets. Many owners do not deny that older primates can be emotionally scarred from their past experiences and lack of a natural upbringing, so they specifically seek out infants who presumably have not suffered such turmoil...yet. A monkey owner identified as "Justine" stated on camera that she wanted a pet that had "already been pulled from its mother." She explained, "Knowing that doesn't make me feel quite as bad. I certainly wouldn't have wanted an adult monkey. All the research, all the other breeders that I've spoken to, all have absolutely said, you know, 'Do not even go there.' So hopefully [with the infant monkey] I have a clean sheet to build that relationship with."[282] As those infant monkey pets grow, many are restrained and modified physically so as to be easier to handle by humans; teeth are removed, males are castrated, and they are bound by leashes and collars whenever they are not enclosed in cages, the better to control their native urges.

Jo Fritz, who originally ran a primate sanctuary called The Primate Foundation of Arizona and later began selling primates back to the laboratories, listed the heartbreaking behavior of the ex-pets she accepted into her sanctuary: ..."the one who came in totally bald because she had nothing to do all day but pull out her own hair; the one that had become completely humanized with her own bed, her own room, and her own clothes. She now lived in a cage because she was jealous of her human brother and severely bit him. The male that had spent eight of his nine years in a laboratory and now does nothing but sit in a corner and rock. There are thirty-two similar stories in the colony."[283] Suffering a range of discomforts, from skin infections to dental and visual problems and even self-afflicted injuries, as well as diabetes and gastrointestinal upset, these primates can be difficult to rehabilitate even after they are forcibly removed from unhealthy conditions.

April Truitt, director of the Primate Rescue Center in Nicholasville, Kentucky, saw perhaps the worst of the worst when five chimpanzees were rescued from Dahlonega, Georgia, in 1998. The chimps had previously been used in circuses, carnivals, and roadside attractions, as well as lived as pets with a previous owner. When the Primate Rescue Center became notified of the chimpanzees, the animals had already spent over twenty years living in underground cages with substandard healthcare and sanitation and rarely had any water or enrichment. Their current owner admitted to being consumed with her own medical conditions and had been unable to pay attention to her pets.[284]

Although the chimpanzees were lucky that the Primate Rescue Center was able to intervene and assume care of the group, even months after their rescue from their feces-encrusted dungeons, the primates still looked like war victims: scarred, misshapen, and mentally unstable, with frequent outbursts and eyes that betrayed the horrors they had lived through.[285] It is fortunate that at the Kentucky sanctuary the five chimpanzees are finally able to live out the rest of their lives with proper food, medical care, fresh air, sunlight, and the social interaction that is so necessary to primate lives in captivity. Most primates living as pets are not so fortunate.

Primates that are surrendered or are removed from pet owners are the rule more than the exception. Sanctuaries turn down numerous requests per week because the demand for primate sanctuary is greater than space and finances can permit. Like zoos, they often can't take any more inhabitants due to space limitations, as well as the complexities of introducing a new individual into an existing social scheme. Health concerns can be a problem, as well, for any contagious diseases can quickly infect the entire troop, and even non-contagious illnesses such as diabetes require a lifetime of intricate care that a facility may be unwilling or unable to provide.

Although most primatologists would prefer to see primates living only in the wild, where they can express natural behaviors, this pipe dream is impractical and simply not possible to support. As imperfect as they are, sanctuaries are sorely needed to supply lifetime care for rescued primates, most of whom will never be able to survive in the wild and some of whom may need to live in captivity for fifty years. Orangutan expert Anne Russon related that "dealing with ex-captive orangutans is like dealing with pollution. We made the mess, so it may be expensive, but it's our fault. So, it seems to me that the onus is on us to try and do what we can to fix up the mess we made. We should be treating them with as much respect as we do other humans."[286]

Kari Bagnall runs Jungle Friends Animal Sanctuary in Florida, a facility with one hundred twenty monkeys (among other primates). She states that their biggest problem is the amount of ex-pets needing a home.[287] Her facility asks pet owners to pay a monthly fee to help provide for the upkeep of their ex-pets once in the care of Jungle Friends.[288]

The physical discomforts suffered by primates living as pets can exacerbate any naturally occurring aggressive tendencies they have, resulting in a cycle of cause and effect. Are they aggressive because they don't feel well due to neglect, or are they neglected because of their unpleasant demeanors? Neither situation is appropriate for an animal; both are deplorable and only serve as reminders of the unsuitability of primates in the pet trade.

Some of the websites that encourage the keeping of primates as pets do also discuss the difficulties involved. Warnings about the social, dietary, and sanitary needs of pet primates are intended to scare away people who think they want a monkey or ape pet on a whim. On such sites, animal welfare groups are generally praised, and animal rights groups tend to be condemned, since if animal rightists had their way, humans would be prohibited from keeping primates as pets. These websites also claim that animal rights groups who advertise the perils of primates as pets distribute lies and scare tactics in an effort to disempower pet owners. Breeding is exalted as a way of perpetuating endangered species. Photographs of adorable monkeys posing with their happy owners are used as examples of a wonderful life possible with a pet primate. Such testimonials often include pictures of monkeys "smiling," when basic research in animal behavior would declare that the poses exhibited in such images are clearly simian expressions of fear.

Back in the early 1800s, a Frenchman named Victor Meunier published *Les Singes Domestiques,* a manifesto supporting a new form of slavery, one that could provide the benefits of free labor without the guilt of imprisoning fellow humans. Meunier's solution to this problem was using apes as slaves. He proposed breeding stations that could be used to develop in apes those skills most appreciated in slaves, such as temperaments docile enough that the safety of the owner could be guaranteed. All the while, the animal kingdom would provide a seemingly endless supply of manual workers. To ensure safety of humans around these slaves, the apes could be castrated and have their canine teeth removed.[289] Although this idea seems ludicrous, if not morally unjustifiable, to most of the world, the concept of using primates in certain fields of work is neither a new one nor obsolete in this day and age.

One area in which primates are used as a type of labor is in the field of service animals, or assistance to the physically disabled. Capuchin monkeys, with their high energy, inquisitive natures, and nimble fingers, can be trained as helping hands to people with physical limitations, such as quadriplegics, or mental limitations, such as agoraphobics. Other primate species, including macaques and chimpanzees, have also been used in the generally unregulated field of service animals. Primates may be trained to complete potentially life-saving tasks, such as administering medication or alerting their owners in case of fire alarms, or they can act as a comfort to someone with debilitating anxiety and social problems.

In the US, the Americans with Disabilities Act stipulates that service animals be allowed to accompany their owners into all public spaces,[290] but as more and more species are being trained as service animals, various government departments are showing concern for the animals' welfare, as well as for the health and safety of the general public. Although there have not yet been any known cases of this occurring from service animals, viruses like Herpes B can be unknowingly present in monkeys and macaques and may be spread to unknowing humans who have contact with them.[291] Because the service animal industry is not highly regulated, humans who request to bring a service animal into a public space are often asked to show what service the animal provides. This can be difficult when service animals help with psychiatric issues, because there is sometimes no tangible way to illustrate that the helper truly does provide a service towards the human.

The helper-monkey field really began in 1989, when the Department of Veterans Affairs approved of the research and plans developed by Boston-based organization Helping Hands. It was at this point that a network of foster homes began preparing monkeys, and disabled people started learning how to live with their new simian companions. Supported through federal grants through 1994, the organization is now fully funded by donations, foundation grants, and workplace giving initiatives. A continuous supply of monkeys is provided through reliance on breeder monkeys that reside at an area zoo. After the capuchins go through extensive training over a seven-year period, they are given to qualified recipients free of charge. The organization supplies continued assistance to the humans who are paired with their trained capuchins, with some pairs staying together for more than twenty years.[292] The capuchins' dexterity and cleverness help to make life easier for people with physical limitations, aiding in tasks as basic as bathing and eating or as complicated as turning on electronic equipment.[293]

The codependence of the disabled humans and their monkeys, combined with the long lifespan of the monkeys, serve the program well, as it takes time and patience to develop a relationship between primate species. The people who live with helper monkeys grow to view them as true family members who not only help make life more comfortable but can also entertain and distract someone living in pain and disability, as well as provide much-needed independence from the help of others humans.[294]

The same issues that plague the primate pet industry also affect the industry providing service monkeys. Quality of care for the animals is a huge issue and something not easily done correctly, and despite their training, the animals can pose a threat to humans from aggressive outbursts and their natural strength and cunning. These animals lead highly unnatural lives as service animals and, as a result, are subject to the same neuroses and behavioral anomalies as primates raised as pets. In fact, helper monkeys' teeth are often removed to ensure there won't be any biting, and males are often neutered, as well.

Trained nonhuman primates benefit not just individuals, but corporations as well. Primates have been used in commercial advertising and entertainment almost since the inception of these industries. Artwork from as far back as the late 1300s depicts captive monkeys, and in more modern times it seems that primates have always been acting in circuses, films, television shows, and commercials. Most people will chuckle at the antics of performing chimpanzees or marvel at their human-like tendencies or abilities to learn physical stunts, but many viewers never consider what led up to that point. That picture-perfect smile and tumbling routine were taught to the primate through repetition and possibly through physical force. It's important to remember that in the wild and in captivity, when nonhuman primates appear to smile, it's a sign of fear. The baring of teeth is a natural reaction to potentially painful stimuli. Be it in response to a prickly food source in a desert range or the threats of a dominant troop member towards a submissive one, showing canines expresses a less-than-savory situation that is causing unrest; conversely, for one species, an open-mouthed macaque stare that shows little of the teeth reveals confidence and a sort of dare from the individual.[295]

It seems that humans are innately drawn towards the behaviors of other primates. They are so similar to us, yet so completely different. They are wild and follow their instincts, but we can train them to act like us—beings who pride ourselves on repression of natural instincts. Using primates in entertainment requires reliance upon a seemingly complete irony: We watch them because we're amazed at their resemblance to humans; yet, we use them in ways that most of us would never use a human, which we justify *because* they're not human.

As primatologists Dale Peterson and Jane Goodall put it, "This kind of ordinary obliviousness about animals in general, once it is applied to monkeys and apes, sees mostly humor. The humor becomes reflexive, automatic, unconscious. An exhausted intellect claws the bottom of the cliché box and retrieves a series of fossilized phrases to express how terribly funny we find our similarity with these creatures to be—'monkey business,' 'money shines,' 'make a monkey out of me,' 'going ape,' 'going bananas,' and so on. The fact is that monkeys and apes make us nervous, and they sometimes make us laugh, because in looking at them we see ourselves. They share, to one degree or another, a shape we regard as the honored shape. Their faces and bodies mirror our faces and bodies. But what we fail to see—or what we see only imperfectly and with some anxiety—is that their minds mirror our minds, as well…the final irony of people laughing at a group of endangered great apes is that great apes are the only animals in the world sophisticated enough, sentient enough, to laugh back."[296]

Apes laugh like humans when unexpected or silly events occur, such as an infant attempting to wrestle with a grown male or the slapstick, bumbling, halfhearted attempts to catch a fleeing individual. Their open mouths, chuckling noises, and relaxed eyes are very similar to signs of humor among human beings. Bonobos, for example, will even make silly faces when bored, much the way human children will, for no reason other than amusement.[297] These behavioral similarities to humans only serve to make primates more endearing and fascinating to us.

Television talk shows are fond of bringing in adorable primates as guests, oftentimes in human clothing, to tumble around and flirt with the audience. Sometimes, when a particular primate character becomes popular, such as Cheetah of the Tarzan movies or Zippy on The David Letterman Show, the character is actually played by a rotating cast of individual primates. It is assumed that the audience would be unable to differentiate between various chimpanzee actors. Most viewers don't understand that the primates they see on television are usually infants, because adults are often too confident and assertive to be trained easily or thoroughly. Some or all of an infant's teeth are typically removed so there is no chance of its biting anyone on set, and electronic shock systems are often hidden under its clothing, so the animal trainers standing just behind the scenes can take control if anything gets out of hand. Faces may even be shaved to make the animals cuter in the eyes of humans.

It should be assumed that any primates seen on television (with the exception of those on documentary shows about animals living in the wild) are highly trained. Primates on television must be trained to perform, because the natural behavior of primates is not particularly funny or marketable. For this reason, animal trainers have developed methods by which to manipulate these strong and highly intelligent animals to do tasks that do not come naturally to them. Animal trainers walk a fine line between emotional bonding with an animal in their custody and creating a sort of emotional assertion over that animal to ensure compliance and protect themselves from retaliation. They often compare the relationship between trainer and trainee to that of a parent and child, in which the right mix of love, patience, and, yes, punishment is supposedly required to bring up a well-behaved charge.

Animal trainers are often highly secretive about their methods, but most involve a mix of psychology and food rewards, although stories abound of what it can take to convince a stubborn 150-pound ape to complete menial tasks and tricks. Knuckle rapping, beatings, and devices such as iron prods and cans of mace have been used to instill fear and enforce submissiveness.[298] Chimpanzees, the species most often used in human entertainment, are naturally independent and curious about testing boundaries, so animal trainers must always be ready to assert their dominance and control in any situation that may arise when training a chimpanzee. It's important to realize that while the American Humane Association monitors the film and television industries to ensure the safety and health of animals used on sets, the organization does not also monitor the living and training conditions of animal actors prior to film days.[299]

There are a few companies that supply most of the primate actors for the entertainment industry in the United States. Some companies have been accused of mistreating and abusing their animal actors, with evidence uncovered mainly via undercover investigations. Abuse within the entertainment world can be difficult to prove on account of the closed-door nature of the industry, and even when abuse is suspected, prosecution can be difficult, as well.

When abuse is suspected, it's not because all animal trainers are intrinsically evil and violent people who enjoy harming animals for no reason. Rather, primate actors may suffer during training simply because, in a way, they are too intelligent for their own good. Primates will not perform intricate routines for basic food rewards. They sometimes reason that even a good reward does not justify doing something they don't want to do. Chimpanzees, like many other primates, are naturally curious, physically powerful, and defiant of the dominance of others, and the only way to ensure their compliance absolutely is with physical threats and injury. Like a human who is in an abusive relationship, a primate who lives under the threat of random violence will slowly but surely develop a more submissive nature and suffer a broken spirit. In that sense, he or she will become more trainable.

The film *Project X,* shot during 1985 and 1986, used the setting of primate research with sign language experiments to explore man's potential to connect with research subjects and recognize their shared genetic similarities, intelligence, and self-awareness. One would think that on a movie shoot for a film that supposedly sympathized with the plight of captive-held primates, the chimpanzee actors would be protected from any sort of abuse or mishandling, but observers described seeing animal trainers beating chimpanzees with clubs, cattle prods, hoses filled with sand and rocks, and their fists. The trainers later described these situations as "knock-down, drag-out fights."[300] As a result, the chimpanzees actors reportedly had a fear of black rubber boots (from being kicked when they attempted escape), and some of them flinched whenever anyone moved suddenly. Fearful of being blacklisted in the entertainment industry, the observers did not report these instances to animal control officers, although accounts claimed that there was once a halt to filming while the trainers were in a state of rage.[301]

When the Los Angeles Department of Animal Regulation decided to investigate what had occurred on the *Project X* location site, six animal trainers were charged with a total of eighteen felony counts of cruelty to animals. Eventually the charges were dropped because the law at that time considered animals solely as objects of ownership. The felony charge of cruelty to animals would only have been applicable if the animals had belonged to someone else, and since Twentieth-Century Fox owned the chimpanzees and hired the trainers, there was no need to press further charges.[302] A separate reinvestigation completed later by the American Humane Association (who had monitored the film shoots in the first place) failed to find proof of animal abuse on the set of *Project X*.[303]

The *Lido de Paris* show at the Stardust Hotel in Las Vegas was a late-1980s explosion of entertaining dancers, visual stunts, and the comic stylings of Bobby Berosini and his troop of five orangutans. The show ran for years, a family-style bill of laughter and grandeur that became a fixture of Las Vegas nightlife. It wasn't until a chorus dancer secretly filmed behind-the-scenes footage of Berosini appearing to beat his orangutans that a scandal broke out. Berosini claimed that the tape was doctored (a claim later refuted by a video expert during the trial) and also that dancers had set him up and had purposely riled his orangutan actors so as to force some disciplinary action that could be recorded.

Both PETA and the Performing Animal Welfare Society (PAWS) received the videotapes and shared them with primate authorities such as Dr. Jane Goodall and Roger Fouts, orangutan expert Dr. Biruté Galdikas, and anthropologist and primatologist H. Lyn Miles, among others, all of whom agreed that inappropriate animal handling was captured on the tape, along with clear signs of distress and fear emanating from the orangutans. A court order allowed veterinarians and a caretaker to examine Berosini's orangutans firsthand, and the animals were found to have wounds and healing lesions and showed typical fearful behavior when Berosini approached, including vocalizing and involuntary urinating and defecating. Their living enclosures (solitary steel boxes that were not appropriate to house such social animals), were one-third the minimum size deemed appropriate by USDA standards.

The various investigations and media reports surrounding Berosini's trial were fraught with emotion and the fear that results when a city built on entertainment is forced to examine one of its most popular entertainers. Conflicts of interests abounded: The president of the Humane Society of Northern Nevada examined the apes and concluded that no signs of abuse were visible, but this same man rented out animals to the entertainment industry on the side and had been promised help with an upcoming telethon if he spoke out on Berosini's behalf. The judge in the trial was a personal friend and past business partner of the head of the financial group backing the Stardust Hotel. Experts that Berosini hired from research labs (the Yerkes Regional Primate Research Center and University of Nevada's Laboratory of Animal Medicine) found nothing wrong with the orangutans' canine teeth removal, their mouths being wired shut, and their submissive, deeply fearful behavior, although they may have been less than sensitive to such issues because their background in research did not focus on the emotional health of captive animals. The media was quick to publicize the positive findings of such inspections (as flawed as they may have been) and reverted to the same kinds of name calling and fear mongering that had plagued animal rightists since the early 1800s, using words such as "militant," "radical," "kooks," and "fanatics" to describe the people who had cared to look into the well-being of animals used for entertainment.

At the time of the alleged abuse, it was not illegal to abuse performing animals in Nevada.[304] In the end, Bobby

Berosini was awarded \$4.2 million in damages,[305] but almost more shocking was the way the atmosphere of the trial prompted denials that the orangutans could suffer. As an editorial piece in the *Las Vegas Review-Journal* exclaimed, "If there are no physical signs of abuse, what's left. [*sic*] Psychological abuse? Of an animal? To assume psychological abuse one must first assume that orangutans *have* psyches."[306] To many concerned people, the assumption that performing animals cannot suffer and do not have psyches was a stretch of the imagination too vast to comprehend.

The animal trainer Sid Yost and his company Amazing Animal Actors had a 20-year history of supplying most of the wild animals used on television, for both movies and major advertising, starting in the 1980's. When primatology student Sarah Baeckler volunteered at Amazing Animal Actors for a year, from 2002-03, she described the treatment and events that she witnessed daily between chimpanzees and their caretakers as "horrifying...sickening acts of emotional, psychological, and physical abuse."[307] Chimpanzees were beaten for misbehaving and for seemingly no reason at all. When Baeckler asked for further direction after being told to use corporal punishment, the following instructions were given to her: "[Hit] hard enough that they know you mean business but not so hard that you do permanent damage...Aim for her head because it's really sturdy...Kick her in the face as hard as you can. You can't hurt her." She was given a hammer to help drive her point across to the chimpanzee actors.[308] Other weapons used included a sawed-off broom handle known as "the ugly stick,"[309] rocks, mallets, and an electric cattle prod.

Baeckler learned that Amazing Animal Actors' methodology was based on little more than fear and intimidation, and the compliance of the chimpanzees was due only to the physical pain that they knew would surely follow any act of defiance. When Baeckler visited other training facilities, she witnessed the same behaviors she had seen exhibited by the Amazing Animal Actors chimps (including threat barks at handlers, and youngsters that appeared timid and hesitant), leading her to surmise that this type of treatment was the norm within the industry. If Amazing Animal Actors had been the only training facility using fear tactics to train their chimpanzees, she assumed, then its chimps would exhibit more submissive behaviors and fear-based neuroses than chimpanzees raised in a more permissive environment. Baeckler's suspicions about the prevalence of violent animal training techniques were later confirmed in privileged conversations with employees at various other exotic animal supply companies.[310]

Publication of Baeckler's investigation findings led to the closure of Amazing Animal Actors, although Yost and his employees denied any wrongdoing and claimed that the animals in their care were always treated lovingly and appropriately.[311] Yost was prohibited from owning or working with primates in the future, and his primates were released to various sanctuaries, although he then started working with a new animal-training venture under the pseudonym "Rick Kelly."[312]

The wildly popular advertising spots on the National Football League's Super Bowl broadcasts have been a common topic of discussion in the United States every February. Although ten commercials featuring great apes have aired during the Super Bowls since 2000, commercials with primate actors are currently being shown less than in previous decades because of a shift in public thought, largely due to the efforts of animal rights groups acting on behalf of nonhuman primates. In response to an ongoing campaign by PETA directed not at the general viewing public but at the advertising agencies themselves (in the hope that when the agency executives are better educated about the realities of primate actors, they will care more about the plight of such performers and decline to use them),[313] as of 2011 eighteen major advertising agencies, including the three largest agencies in the United States—McCann Erickson, BBDO, and Young & Rubicam—have committed to no longer use great apes in their work, and the number continues to grow.[314] Lobbying has also resulted in some major corporations, including Pfizer, Dodge, and Travelers Insurance, ending existing campaigns that used primate actors.[315]

With each ad agency and corporation that makes the move away from using nonhuman primate actors to hawk various wares, a gradual shift occurs. The public is growing increasingly uncomfortable with seeing such animal actors, causing this type of exploitation no longer to be appreciated as a viable economic tool. Today, it's not difficult to envision a time in the future when there will be no nonhuman primates used in advertising at all.

Using primates in entertainment occurs not just in Hollywood and not just in the United States. Tourist traps in foreign locales where primates live naturally often utilize captive monkeys and apes who perform dances or pose for photos...all for a price, of course. For instance, in the slums of Jakarta, Indonesia, chained and bound macaques are forced to perform for mere coins. A person versed in the natural behaviors of macaques would easily recognize that the animals are petrified and crying out, having no choice but to perform or suffer physical punishment. Most of these macaques were taken as infants from their mothers by poachers and denied food and freedom of movement until they learned the skills, such as standing upright or dancing, that their owners demanded. Housed individually in wooden crates and denied the types of social activities that would dominate their life in the wild, these macaques live their entire squalid lives mentally disturbed and unhappy. Most tourists that pose for photos in front of the trained animals are oblivious to their suffering.[316]

Using nonhuman primates in entertainment has proved to be a relatively easy and reliable method of getting people's attention, but creating and maintaining these primate entertainers long-term is anything but simple. Many primate actors are chimpanzees or orangutans (both great apes). Consider the natural life cycle of a typical ape: In the wild, apes will nurse for up to six years, but in order to be entertainers, apes must ideally be less than eight years old, for after that age their maturity and strength make them too dangerous and unpredictable to be handled safely. This means that ape actors are denied normal childhoods and bonding rituals with their mothers and other conspecifics so that they may spend just a few years working at a job they will eventually age out of.

Patti Ragan, founder and director of the Center for Great Apes in Wauchula, Florida, estimates that annual care for a retired ape actor is approximately $20,000. Using as an example a popular commercial with four chimpanzees, such as a CareerBuilder.com ad that ran during the 2011 Super Bowl, and considering that those four chimpanzees could live for half a century, the cost of their retirement after age eight totals $3.4 million...all for the sake of one commercial.[317] Sadly, the companies that hire primate actors for these jobs do not contribute to their care later in life, nor do the breeders that brought them into the world, nor do the trainers who were in complete control of the primates' lives for those working years. The sanctuary system is saddled with years and years of costly care for primate actors who deserve more attention than is available to them.

"It looks to me like these commercials are making these animals seem cute and perfectly well-cared for," said Barbara J. King, anthropology professor at the College of William and Mary. "It's not clear to me from the surface of it why consumers would necessarily be concerned unless someone tells them the back story."[318] When people neglect to associate an endangered primate or a cash-strapped primate sanctuary with the zany baby chimps they see frolicking on TV, they are less likely to contribute financially for their long-term care. People typically don't understand that nonhuman primates need help if they look adorable and lovable and seem to be happy playing around while shilling merchandise.

Primates who are lucky enough to be rescued from lives spent in laboratories or in the entertainment industry can be discovered to have frightening provenance records. Oftentimes, sanctuaries may not know what studies their inhabitants were involved in, whether they were wild-born or not, or what personal traumas they may have suffered in the past; however, clues often arise once they start their new lives in a sanctuary.

A chimpanzee that was used as a lab breeder and had all four of her babies removed from her spent the later part of her years collecting dolls at Washington's Chimpanzee Sanctuary Northwest, mothering the inanimate objects in ways she was never allowed to nurture her own offspring.[319] Other chimpanzees show propensities to wear human clothing and jewelry, or use utensils to eat food, giving clues that they were raised in a human family and learned the human ways to do things. Even if the surrounding humans don't always know what happened to these animals, it's important to remember that *the primates do*. Their past is as hardwired into their brains as human memory is, helping primates to efficiently recall both the good and the bad experiences of their lives.

When considering the history of performing primates, Victor Meunier's theory of ape slavery no longer seems so strange. As Dale Peterson states in *Visions of Caliban*, "Slavery is not defined by action, but by condition." [320] Humans would typically describe as slaves those who are mandated to work and who cannot escape the existence forced upon them. That most people assume that the term *slavery* is unique to the human condition doesn't make it so. When rational beings are forced to spend their lives serving others, working without pay and against their free will, that must be considered tantamount to slavery, and something is indeed very wrong with that practice.

4. Bioresearch

Research procedures should not be sustained merely because they have been used in the past.[321]

- National Research Council

Previous chapters of this book have noted that vivisection, or experimentation on live animals, has been a hotly debated issue throughout history, practically since the very inception of such procedures. There is no doubt that the human race has benefited from animal research. Modern medications have allowed people to live longer and control disease more successfully. Advancements in cleaners have permitted people to scrub and sterilize their environments more easily and thoroughly, and cosmetics can now be used without fear of infection or toxicity. But was all this production necessary?

Any trip to a local big-box retailer reveals aisles of proprietary detergents, rows and rows of nail polishes, and more versions of pain relievers than there are types of pain. Are the lives of thousands of animals worth this plethora of choices of commercial goods? It may be easy to dismiss duplicative retail products, but emotions can get charged more easily about the necessity of medication developed through animal testing.

We all know people whose lives have been greatly extended and improved by medications and surgical devices. Virtually all these developments have at some point earned FDA approval through research involving live animals. Although the FDA does not mandate animal testing for all new products brought to market,[322] individual corporations may decide to have products tested on animals to ensure safety before presenting them for human use or consumption. There are few prohibitions in place in regard to the extent or inherent usefulness of such product tests.

While nonhuman primates account for less than 25 percent of all animals used in laboratory tests (the majority of which are rodents), there are approximately 30 different primate species currently used in testing.[323] Chimpanzees and marmosets are popular subjects of research for Hepatitis A, B, and C vaccines, and marmosets were also instrumental in researching vaccines for varicella zoster (chicken pox). Influenza B research often involved rhesus macaques, and HIV/AIDS research typically used chimpanzees as a model. The cure for polio was found through research conducted on rhesus macaques as well, which, studied alongside baboons, were also instrumental in finding a vaccine against measles, mumps, and rubella.[324]

In earlier days of animal research rhesus macaques were most commonly studied, along with some capuchins and squirrel monkeys. Advances in genetics advocated for the use of chimpanzees, which became popular modern research subjects because of their close genetic relationship with humans and their more permissive "threatened" status, as defined by the U.S. Fish and Wildlife Service (other great apes, such a gorillas and orangutans, are listed as "endangered," which makes them less accessible for research purposes). Tattooed with serial numbers for identification, chimpanzees are transformed from individuals with their own distinct personalities into cogs in the great scientific machine known as animal research. As public awareness has been raised about chimpanzees used in research, and as scientific advances have allowed for alternative methods of study, chimpanzees are no longer used in as many laboratories as they once were. In fact, the United States and the African country of Gabon are the only countries where chimpanzees are still use in bioresearch.[325]

There are well-intentioned documents that provide recommendations and advice on how to care for and maintain healthy primates properly in a research setting, such as *Guide for the Care and Use of Laboratory Animals* (National Research Council, 1996), reinforced by *U.S. Government Principles for the Utilization and Care of Vertebrate Animals Used in Testing, Research, and Training* (Office of Science and Technology

Policy, 1985). As previously discussed, the Animal Welfare Act (AWA) only sets standards for the procurement and living conditions of animals, not the actual experiences they may undergo as research subjects.

Although other countries have committees assigned to consider the ethics of animal testing, some of which have the ability to approve or deny the commencement of testing, the United States' animal testing system permits much freedom for the researcher. While some organizations are required to report when animals are subjected to pain without pain-relieving drugs, this stipulation doesn't apply to government agencies, and the purposes of experiments are never questioned. All a researcher must do to justify causing research subjects pain without providing pain relievers is to state that the presence of pain relievers would negatively affect the results of the test, and then go on as usual,[326] regardless of the veracity of this claim.

Institutionalized primate research as we know it today began in 1923 when psychobiologist Robert Yerkes purchased three pairs of chimpanzees. Two died quickly, but the remaining four formed the base of his Florida laboratory. With funding and support first from Yale University and then Emory University, what grew into the Yerkes Primate Research Center is now located in Georgia and has spawned countless other facilities that do bioresearch involving numerous nonhuman primate species, including six federally funded National Institutes of Health (NIH) primate research centers opened in the 1960s. The NIH developed the Chimpanzee Breeding and Research Program in 1986 to ensure ample subjects for the burgeoning field of HIV/AIDS research; nine years later the NIH stated that there was a surplus of chimpanzees for research and halted further breeding.[327]

Protesting against using animals in research is a mainstay of animal rightists, but primates used in research can sometimes elicit just as passionate a response from animal welfarists. Their reasons are often the same reasons that primates are so frequently used in research: They are similar to humans. Their self-awareness and ability to communicate with humans is not found elsewhere in the animal kingdom.

Nonhuman primate species are often quite large and, like humans, they require regular physical and mental exercise for health, which cages and sterile environments simply can't provide. They may also require highly specific environments to fulfill their natural urges, and labs just can't replicate those. After all, it's hardly possible for a gibbon to brachiate hand-over-hand in a laboratory the way one would on jungle vines.

Perhaps one of the cruelest aspects of primates' being used in research involves restriction of their highly sociable nature. Most primates live in groups with defined social expectations, practices, and behaviors that serve to both calm and placate the group and the individuals living within it. Irwin Bernstein, a University of Georgia psychologist and chair of the Committee on Well-Being of Nonhuman Primates, which in 1998 authored *The Psychological Well-Being of Nonhuman Primates* in response to the 1985 amendment to the Animal Welfare Act, flatly stated, "The fact is, primates are social. It's not just that people are social, it's pervasive through the primate order. That's how they cope, with social support. You can give them all the toys in the world and it's not going to be a substitute for companionship. And biomedical scientists who say, 'Oh, but the monkeys can see and hear each other,' are ignoring the fact that a monkey's primary social contact is physical contact. They need to touch."[328]

The document produced by Bernstein's committee, an all-encompassing guide to the appropriate care for primates living in captivity, states objectively that "social housing is a critical component of psychological well-being.... Although a social living situation is important, there can be practical and scientific reasons for using individual housing, such as research protocols, medical conditions, the possibility of disease transmission, hyperaggressiveness, and hypersubmissiveness. When experimental protocols require individual housing, nonhuman primates should, whenever it is possible, have visual, auditory, or olfactory contact with each other."[329]

Because of the procedures involved in maintaining accurate records and tests, most research protocol requires that subjects be housed singly to avoid contamination or confusion of sampling. The side effect of such isolation is not temporary for many research subjects; it can affect the quality of their remaining years. After reviewing the Alamagordo Primate Facility in New Mexico, a memo from the USDA noted that "because of the long-term housing in these single cages, these chimpanzees have not been able to perform species-specific behaviors, including social and physical behaviors... normal exercise and full stretching are not possible in these cages. This has...caused physical and psychological suffering and distress to the chimpanzees."[330] Despite the well-known side effects of species-inappropriate living and housing conditions, the majority of the animal testing community does not seem to think these concerns are of enough value to question or change the procedures causing such distress.

Because there is such discord among animal rightists and welfarists and vivisectors, animal testing labs have become very closed, insular communities that do their best to keep dissenters away. Every so often, however, a company will make the news due to an undercover investigation by organizations that reveal blatant suffering of animals undergoing research, proving that the systems in place to protect research animals have flaws and holes.

The organization Stop Animal Exploitation Now (SAEN) is active in publicizing what they view as atrocities in the animal research community, be it in for-profit laboratories or state university systems. SAEN is one of many like-minded groups working to patch what they perceive to be holes in the animal testing system. Their website explains how they work at "exposing the truth to wipe out animal experimentation"[331] via volunteer and grassroots investigations into facilities across the United States. Their reports are submitted to the USDA and commonly list unexpected deaths of research subjects, oftentimes primates, from unnatural causes stemming from negligence or human error or irresponsible laboratory procedures. As founder

Michael Budkie explained in an article in Wisconsin's *The Cap Times*, "The fact is, these deaths are happening literally across the United States and the USDA is doing virtually nothing about it. So what's the point of even having an Animal Welfare Act?"[332]

Animal testing is often viewed as a necessary component of modern medicine and scientific advancement. If animal testing were suddenly abolished, the development of new medications would be slowed, and products might have a higher risk of unstudied side effects. It's understandable that many people would not be satisfied with either of these two choices; for this reason, the more extreme animal welfarists and rightists that aim to end all animal testing tend to have very little luck getting public support. Certainly innovation would not be halted completely if animal testing were to end, as computerized testing is becoming more and more plausible as a complete replacement for vivisection, and there is still the highly debatable but potentially feasible option of using only human test subjects, but progress would definitely slow during any transition. Oftentimes, it's more expedient for activists to focus on improving regulations and on limitations to animal use in testing, since a minor victory on behalf of animals may be preferable to no change at all.

It's important to acknowledge that many people involved in animal research have no ethical problems with laboratory testing of live animals. Frederick King, onetime director of the Yerkes Regional Primate Research Center, explained why primates specifically are so often chosen for vivisection: "[S]imilarities in the biological mechanisms of humans and primates underlie the value of these animals for research in a broad range of disciplines…. The complexity of the primate brain and its similarity to that of humans makes [sic] primates excellent subjects for the study of motivational states such as hunger, thirst, and emotion…Because the primate brain shares with humans a high degree of plasticity, their cognitive and social behaviors are heavily dependent on learning and the environment, as is the human behavioral repertoire…[P]rimates in general develop socially and relate to each other and their environments in ways that are more similar to humans than to other animals."[333]

Vivisection has been defended under the First Amendment as necessary to the development of new ideas.[334] In general, Americans believe in their right to every possible medical advancement, and that depends upon much research. There are currently limits on how such research may be conducted, such as prohibitions against bioresearch performed on unknowing or unwilling human subjects; but animal experimentation is considered more permissible than similar experimentation on humans because many people believe either that animals are not capable of suffering to the degree that humans can suffer or that any animal suffering experienced as a result of the research is justified by the potential for scientific progress that can benefit humans. After all, many human diseases have been eradicated and ominous product side effects diminished due to bioresearch involving primates.

For people who value human concerns over all others, there is no moral quandary regarding animal research. Additionally, researchers are quick to discuss the excellent veterinary care that primates receive in laboratories, and some even point to enrichment program features, such as videos that research subjects can watch or decorations that liven up laboratory walls. Public perception of animal research, like any other hotly discussed issue, can easily be swayed by discretion and careful publicity coming from what are considered reputable scientific resources.

The biological and genetic proximity of human and nonhuman primates means that their reactions to diseases and immunizations are often similar, though not always so. Some diseases, such as herpes B and HIV, can live in the bodies of nonhuman primates without any attendant suffering of the host, whereas the same diseases can be devastating and often fatal to an infected human. Bioresearch is particularly helpful when scientists aim to learn about aspects of an illness that are difficult to study when it occurs naturally in the human population. For example, an untested human can be infected with HIV for months without knowing it. If scientists want to learn about the early onset of HIV in un-medicated people, they have almost no way to do so (short of using human volunteers) other than to infect nonhuman research subjects purposely in a controlled environment and monitor the progression of the disease in the hope that the findings may be useful in human cases, as well.[335]

The grant money and the prestige so often intertwined with bioresearch encourage researchers to continue with and not limit their animal testing, regardless of their personal beliefs about why they do animal testing in the first place. There is little incentive for scientists to avoid animal testing. When the U.S. Fish and Wildlife Service was considering changing the status of chimpanzees from threatened to endangered in the late 1980s in order to recognize and document an increasingly imperiled species, they received over 54,000 letters in support of the action and only nine opposing the change. Eight of those nine letters were from federally funded researchers who complained that the change in status would negatively affect their ability to continue research on captive chimpanzees. The one remaining letter was from a circus. The letters in opposition proposed listing wild chimpanzees as endangered but listing captive chimpanzees as only threatened. Despite the protests of primatologist Roger Fouts, PETA, and other animal defenders, and in an effort to not further complicate the procedures of animal researchers, the government agency agreed to the researchers' proposed use of the two separate classifications for captive and wild chimpanzees.[336]

The disappointing realization that the U.S. Fish and Wildlife Service would value the opinions of eight scientists over almost 54,000 civilians, combined with the scientists' blatant self-interest trumping concern for an environmental cause, was almost as upsetting to activists as the knowledge that captive chimpanzees could not enjoy the same protections as their free wild brethren. Whether the scientists' reactions represented an intentional and convenient turning of a blind eye to ethical considerations about animals or an honest expression of their beliefs remains unknown.

One case of convenient ignorance of animal suffering took place in 1984 at the Head Injury Clinical Research Laboratory at the University of Pennsylvania, which was using baboons to investigate cranial trauma when the animal rights group Animal Liberation Front (ALF) infiltrated the facility. ALF made off with videotapes of experimental procedures showing conscious baboons undergoing brain surgery and other head trauma (including acceleration experiments designed to observe a baboon's head pressed against stone helmets at a force as high as two thousand times that of gravity),[337] in addition to other inhumane behaviors by the researcher staff. After PETA edited the tapes and distributed the video to media outlets worldwide, public outrage grew with the knowledge that federal funds had paid for the horrific experiments.

Despite the lab's defense that anesthesia had been used on the baboons, lab records and neutral experts confirmed that the levels of anesthesia were insufficient, at best, and that it was often not used at all. Dr. Thomas Gennarelli, one of the two directors of the clinic, could not provide proof of any instance of the experiments' results actually being used in a practical application.[338] The suffering of so many baboons was for nothing, and, to add insult to literal injury, the lab had by that time received millions of dollars worth of federal funding to pay for nothing but that suffering.

In response, the university performed its own investigation, and after its committee neglected to find any wrongdoing within the lab, the National Institutes of Health (which dispenses federal funding to researchers) failed to acknowledge the obvious conflict of interest inherent in UPenn's evaluating its own staff, supported the lack of findings by UPenn's committee, and decided to continue funding. Three days of public protest in July of 1985 finally convinced the lab to reconsider, and on July 18, 1985, the Head Injury Clinical Research Laboratory was closed and the research halted indefinitely.[339]

As animal law expert Gary Francione commented, "The Gennarelli case stands as probably the most important animal rights case of this century. Well-respected researchers at a well-respected university were depicted on videotape—that they had taken themselves—as engaged in what could only be described as brutal and barbaric behavior. Moreover, the researchers, the university, and the government all exerted extraordinary effort to deny what was clearly and unambiguously shown in the videotapes and contained in the researchers' own writings."[340] Instances such as this illustrate the inherent flaws in self-regulating industries without watchdog interference, the ambiguity of some animal welfare laws, and the reluctance of the federal government to question research that it had previously supported staunchly, both financially and figuratively.

Primate research is not limited to medication development, retail product safety, and general academic study. The United States Armed Forces has long been using primates to test agents and methods of warfare. For example, the Primate Equilibrium Platform (PEP) was an airplane simulator at Brooks Air Force Base in Texas where rhesus macaques were trained to "fly" an airplane with a synthetic program that involved using a joy stick. As using a joy stick to maintain equilibrium is not something that rhesus macaques normally do, training in the PEP system included the use of restraints, electric shocks, food deprivation (and raisin rewards for maximum efficiency), and violent movements of the platform, all scheduled for weeks at a time. It was only after the individual subjects had proven that they had learned how to fly that the actual research began. After exposure to radiation and chemical warfare agents, the monkeys' ability to fly was tested. Often sick, in pain, and disoriented, the small monkeys were put into the PEP system and forced to perform under the influence of toxins that could have killed a full-grown human.[341]

Neil Armstrong may have made that first giant leap for mankind, but before then many primates had bravely gone where no man would go at all. The 1960s ushered in a new era of space exploration for the human race, and the United States government was careful to test proposed procedures and treatments on living things such as seeds, insects, rodents, and, finally, primates before allowing humans to utilize any of the developments. Although records are lacking from the beginning of primate testing involving spacecraft, perhaps due to the fact that people at the time thought the research was unimportant or unworthy of recording meticulously, it's clear that nonhuman primates played an important role in this part of scientific history.

White Sands Missile Range and Holloman Air Force Base are both situated in the desert areas around Alamogordo, New Mexico. Here, monkeys and chimpanzees were relied upon in procedural testing because of their anatomical similarity to humans, as well as their manual dexterity and ability to learn and perform tasks. Slowly acclimated to the seats, equipment, and sensations they would be forced to experience in the missile capsules, the primates were taught compliance through food rewards. This systematic training continued, according to one researcher, until the primates could sit in the chairs "indefinitely."[342]

Once the subjects learned to perform a more complex task, such as a sequence of lever pushes, they would be required to complete that task under different and increasingly difficult circumstances, such as gravity loss in rotating cylinders, among other physical stressors. Video footage from this time shows scared and hesitant chimpanzees clinging to researchers, reluctant to put on gear and sit in foreboding looking experimental chairs, perhaps because they remembered negative reinforcement such as shocks and electrodes or perhaps simply because it all seemed so alien to beings who, not too long before, had been ripped away from their childhood and natural homes.

Primates were first used in these programs at Holloman Air Force Base. Monkeys tested rocket flights in four missions in an effort to determine the effect of g-forces on a living being as it left and reentered Earth's atmosphere. All the subjects died as a result of either faulty parachutes or exposure after being lost in the desert after a landing.[343]

Chimpanzees were used in testing at White Sands Missile Range and blasted off in various contraptions at high speeds of 300-400 miles per hour, then stopped suddenly with brakes so that scientists could examine what would happen to their internal organs. Such experiments proved correct the hypothesis that sharp deceleration at such high speeds causes the brain to hit the sides of the skull. It stops moving only after the body does, resulting in sudden death.[344]

The first successful space experiment occurred on May 21, 1952, when two rhesus macaques and some mice were launched 26 kilometers into the atmosphere and survived the trip. The media lauded the project a success, proudly deeming the macaques "physically unharmed and in jolly good spirits; the first creatures from outer space in all history."[345] America's exploration into space was ramped up by the success of that launch.

President John F. Kennedy's space program legacy began in 1958, soon after the creation of the National Aeronautics and Space Administration (NASA). By May of 1959, two monkeys were successfully launched into space on a Jupiter missile suborbital flight and returned alive after splashdown.[346] In December of that year, a rhesus monkey named SAM (named for the Air Force School of Aviation Medicine) flew 53 miles high as part of the Project Mercury program.[347]

Meanwhile, chimpanzees were being brought to the United States from Cameroon to fill research slots at Holloman Air Force Base. Although records are vague, it has been reported that the methods used to obtain the young chimpanzees were less than ethical and likely involved killing of the mothers in order to take the infants. At Holloman they completed training regimens that at the least bordered on physical and mental abuse, involving many hours of repetitive tasks such as flicking switches and sitting in uncomfortable chairs, activities alien to a physically active wild animal.

One of the chimpanzees from Cameroon was fated to make history, although whether he could be considered lucky or not is debatable. His name was HAM, an acronym for Holloman Aero-Medical, and he was a three-year-old male chimp who was chosen to participate on account of his intelligence and affable, loving nature. HAM was trained with food rewards to complete certain tasks, and his success in that ended with a launch into space from Cape Canaveral in Florida on January 31, 1961.[348] Mechanical difficulties resulted in his rocket burning fuel faster than expected and flying 130 miles farther than expected, at speeds up to 16 g's. His capsule overheated and landed harder than expected, which caused it to take on water.[349]

After a successful rescue, HAM amazingly showed no obvious ill effects from his adventure (other than an understandable refusal to sit in his flight seat again) and became a media darling; however, chimpanzee expert Dr. Jane Goodall has said that the wildly popular photographs of HAM smiling after his flight reflect true fear. In fact, she claims that his face shows the most fear she's ever seen exhibited by a chimpanzee.[350] When doubtful viewers pointed out that HAM's smile might indeed have been from fear, the space program staged a public event featuring HAM and the capsule in which he had ridden into space. When the big event started, much to NASA's chagrin, HAM refused to go back into the capsule, and even four grown men could not overpower and force him to sit in it.[351] Five months later, astronaut Alan Shepard replicated HAM's flight successfully.

After HAM came Enos, a chimpanzee chosen to test a new type of missile. A five-year-old with a bit of a wild streak, he was smart and daring and was chosen to make a three-orbit flight on a Mercury-Atlas rocket on November 29, 1961. Enos's independent nature meant that he often had to be restrained with straps and leashes during the experimental procedures, and one of the researchers compared the gear used to contain Enos to a straitjacket.[352] After the second orbit, he was brought back to Earth, reportedly due to equipment malfunctions including one in which he was being electrically shocked instead of rewarded with food after every correct action he made aboard the craft. Despite the repeated punishments, Enos continued to perform the correct actions aboard the spacecraft. His three-hour flight involved 181 minutes without gravity and ended with his rescue after a scheduled splashdown in the Atlantic.[353] Three months later, John Glenn would orbit the earth and return to wide acclaim, but it's been reported that upon meeting the president's daughter at one of his many celebrations, Glenn was asked by the young Caroline Kennedy, "Where's the monkey?"[354]

At the time of HAM's and Enos's NASA voyages, the Cold War (the ongoing tension that propelled an arms race and a space race between the United States and its allies and what was then the Soviet Union and its allies) was at the forefront of American media and politics. For many obvious reasons, the United States wanted to be the first to put men in space, and the use of primates in the space program was initiated to ensure safety for eventual human astronauts. Ironically, it was primate testing that kept the US from its goal. HAM's flight occurred in January of 1961, and while NASA was still completing testing in April of that year, in preparation for its first human flight, the Soviets became the first nation to put a man in space.[355]

In September of 1962, President Kennedy gave a memorable speech, rallying America in support of the race to the moon and exploration of the great space beyond. "In short," he postulated, "our leadership in science and in industry, our hopes for peace and security, our obligations to ourselves as well as others, all require us to make this effort, to solve these mysteries, to solve them for the good of all men...."[356]

In combination with those "obligations to ourselves and others,"[357] America's scientific advancement relied not only on the work and sacrifices of human beings to claim our stake in space, it relied on the bravery and sacrifices of nonhuman primates, as well. President Kennedy proclaimed the early space missions as "the most hazardous and dangerous and greatest adventure on which man has ever embarked,"[358] and he said this before man had ever actually begun it. Nonhuman primates had already embarked and suffered and survived, all to ensure the future safety of their fellow primates, human beings.

It would be heartening to hear that HAM, Enos, and the macaques that were the first aeronautics testers were treated like royalty for the rest of their days, since they should really have been hailed as heroes, much like the human astronauts who followed in their footsteps. Alas, that is not the case. HAM ended up on display at the National Zoological Park in Washington, DC in 1963, and although his celebrity status surely brought some attention to the zoo, he was far too acclimated to living among humans to be able to socialize with the other chimpanzees already in residence at the zoo. In the 1980s he was moved to the North Carolina State Zoo and was finally able to join a chimpanzee social group, but he died in 1983 at the young age of 26.[359] After much debate, the Air Force and the Smithsonian Institute kept his skeleton for research, and his other remains were buried at the New Mexico Museum of Space.[360] Enos died of dysentery (an infection said to be unrelated to his space flight) only a few months after his mission, and there are no records that note the disposition of his remains.[361] It is presumed that they were simply discarded.

The remaining chimpanzees held at the Air Force testing base were no longer needed after the 1960s, and starting in the 1970s they were leased out to medical labs for more research. In 1997, the Air Force declared that its chimpanzees were considered surplus and a financial burden, and they would be given away permanently for medical research to universities in New York and New Mexico, as well as to the Coulston Foundation, a facility with a history of abuse of chimpanzees and three prior FDA violations.[362] It was at this point that primatologist Dr. Carole Noon became upset at the irony of the way the human astronauts were feted and celebrated as true American heroes, while the 141 remaining nonhuman astronauts were "re-assigned to a hazardous research environment."[363] Thirty chimps were retired to Primarily Primates, a sanctuary in Texas, and Dr. Noon successfully sued the Air Force to obtain custody of the remaining chimpanzees. To house the new residents, she built the sanctuary Save the Chimps (then called the Center for Captive Chimpanzee Care), in order to provide lifelong refuge to the chimpanzees that had lived through so much already.[364]

Chimpanzees were not the only primates involved in the space race. Additional NASA testing involved rhesus macaques, very social and intensely emotional primates that share 92 percent of the same genes with humans.[365] They had performed well in previous testing experiences, showing very little motion sickness, and they had innate, hardy survival skills that had helped them thrive in modern cities throughout Asia and India. Despite research that rhesus macaques were incapable of comprehending anything they could not touch with their hands, they proved to have excellent dexterity and understanding of joystick-controlled video games, to the point where they would not only perform for food rewards but would play for their own enjoyment and curiosity. The video game research was conducted by Duane Rumbaugh, who went on to prove that not only were rhesus macaques able to manipulate the graphics on a computer screen accurately, but they also could use computer programs to understand the relationships between numbers.[366]

The typically underestimated intelligence of rhesus macaques (now the most widely used primates in research laboratories) raises questions about the ethics of keeping them confined and mentally unstimulated for years or for an entire lifetime. Test results like those of Rumbaugh, which described a high level of intelligence and comprehension in rhesus macaques, could contribute to the end of their use in medical research, in which case scientists would be hard pressed to find research replacements with comparable aptitudes.

In 1981, an incident that some believe kick-started the combative protests against laboratory testing occurred when a college student named Alex Pacheco began volunteering at The Institute for Behavioral Research, a small, federally funded primate research facility in Silver Spring, Maryland. During the previous year, Alex had joined with Ingrid Newkirk, another activist, to form what would become People for the Ethical Treatment of Animals, or PETA. When he volunteered at the Institute, he claimed to have a passion for the research and did not tell his coworkers that he was active in the animal rights movement.

Alex began taking notes and photographs, even smuggling in animal care experts, such as veterinarians, in order to document accurately the substandard care within the lab. He observed that seventeen macaques were not only being subjected to cruel experimentation involving spinal surgery, the binding and numbing of limbs, and the use of electric shocks, but that roaches and rat droppings and urine abounded in the building, and the macaques there showed clear signs of severe distress, including obsessive-compulsive behavior and self-mutilation to the point of infection and self-amputation of limbs.[367]

The resulting police investigation led to charges of animal cruelty against Edward Taub, the chief research scientist at The Institute for Behavioral Research, although all charges were eventually dismissed due to Taub's research satisfying the definition of what was deemed within "the normal bounds of science."[368] At the time, some states exempted federally funded laboratories from complying with state animal control and cruelty laws.[369] Although this is just one case of many that have since hit the news, and not one with the most shocking conditions revealed, it was pivotal in demonstrating to scientists that animal activists meant business and had no qualms about resorting to deception in order to protect the animals involved.[370]

In 1986, SEMA, a laboratory in Rockville, Maryland, was infiltrated by the animal rights group True Friends. Activists broke into the facility and documented the conditions of approximately 500 primates housed there for AIDS research. The video they shot showed substandard conditions that many people deemed cruel. Apes and monkeys were kept in isolettes (small, bare, cage-type enclosures where laboratory animals are often housed in areas too small for them to move naturally). The isolettes used at SEMA were only 40 inches high and 26 inches wide, and the primates inside them reflected severe mental and physical deprivation and suffering through behaviors such as continuous rocking and blank staring.[371]

PETA got involved and dispersed the video to the media and other animal advocates to illustrate that, although the SEMA lab met the standards of care established by the National Institutes of Health, these were neither sufficient as protection nor empathetic to the animals involved. Although major figures in the world of primate care, such as Jane Goodall and Roger

Fouts, got involved in the fight to improve conditions for primates in laboratories, the SEMA lab continued operations and only slightly improved living conditions for its primates by enlarging and decorating its isolettes.[372]

Although some bioresearch has been proven to be cruel to the animals involved, not all primate research involves medication or physically painful procedures. Due to their high level of intelligence and awareness, as well as the similarities with humans in response to certain stimuli, primates are often the subjects of psychological testing.

One of the most well-known (and hotly debated) primate researchers was Harry Harlow. In the 1960s, while working at the Primate Research Center in Madison, Wisconsin, Harlow published results of his psychological studies involving social isolation, psychopathy, and abusive mother figures. By raising infant primates in distressing situations—which, depending on the experiment, involved years of having no contact with another living creature, or interacting with "mother" substitutes that were in essence cloth-covered machines designed to physically harm the infant once imprinting (attachment to a mother figure) had occurred—Harlow proclaimed to have discovered the arguably logical fact that infants will cling to mother figures despite also receiving pain from them and that infants raised in social isolation develop pathological tendencies and are unable to form normal relationships later in life. Such socially deprived female primates were then forcibly impregnated and observed as they ignored, mutilated, and killed their subsequent offspring. As described by author Deborah Blum, "His approach was very direct: the easiest way to investigate a loving heart is to break it; the shortest cut to explore a relationship is to sever it."[373]

Harlow's work proved the importance of touch and social bonding through the deprivation of those very elements that are most important to developing a healthy psyche. As awful and unnecessary as Harlow's dark research concepts may seem to many people (they've been described as "sadism, cloaked in jargon"[374]), his experiments did not die with him. Further deprivation, depression, and isolation tests have been carried out in labs and universities, and it's uncertain whether information gleaned from the suffering has ever been put to any practical use.

Harlow's experiments were hardly the most gruesome, especially in comparison with some of the physical abuses inflicted in other laboratories. Primates have been used to test lethal doses of radiation, nuclear bombs, electric shocks, the impact of gunshots at close range, and spinal cord injuries, and a cornucopia of fatal and painful diseases have intentionally been introduced into their bodies, just to observe what happens as they are sickened and injured.[375]

The usefulness of applying animal bioresearch results as models for human medicine has been questioned because in many ways animals react differently to treatments than humans do. Some results of animal testing that were thought to be applicable to humans proved to be extremely dangerous to human health, partly because humans are physiologically different from animal subjects in key areas. Many medications prescribed to humans, such as insulin, morphine, and penicillin, are toxic or damaging to certain animals often used in test labs, such as rabbits, mice, and guinea pigs. The American Medical Association has admitted that animal testing "frequently prove(s) little or nothing...to correlate to humans."[376]

It's difficult not to notice that, for the most part, chimpanzees seem to have played the largest, most visible role in primate testing (though there are notable exceptions, such as Harlow's monkeys and the macaques used in Air Force tests). As of May 2011, there were 937 chimpanzees being studied in research facilities, 436 of them supported by federal funding.[377] Although various macaques (crab-eating, rhesus, and pig-tailed), monkeys (squirrel, vervet, and green), and capuchins are often imported into the US for experimentation,[378] over time, chimpanzees have been studied and observed more than any other ape. Primatologist Benjamin Beck created the term "chimpocentrism" to signify the imbalance in knowledge gleaned from the contributions of *pan troglodytes*.[379] Many other primatologists, anthropologists, and biologists agree that it can be dangerous and scientifically irresponsible to apply knowledge of chimpanzee physiology and psychology to the diverse range of non-chimpanzee primates.

Chimpocentrism can be clearly seen in the HIV/AIDS studies of the 1980s through the date of this writing, as researchers have tried in vain to find a cure using humankind's closest relative. Although primates and humans were shown to react similarly to major health threats such as polio, rubella, corneal vision problems, hepatitis B, and thalidomide birth defects, Robert Gallo, an American pioneer in identifying the HIV/AIDS virus, has supported increased research in human beings and less on chimpanzees (which had been the standard "guinea pig" for early HIV/AIDS research). Gay activists such as Larry Kramer have actually requested to be subjects in HIV/AIDS research, believing that this practice would produce more usable data and future vaccines than continued primate research.[380]

Chimpanzees are the nonhuman primate most reactive to the HIV virus, but it's important to note that they don't get sick from it the same way human beings do. A decade-long ongoing fecal study by Beatrice Hahn of the University of Alabama at Birmingham found that the mortality of chimpanzees at Jane Goodall's research site at Gombe, Tanzania, may be affected by simian immunodeficiency virus infection (SIVcpz), the disease that is believed to have been the biological origin of HIV and AIDS. Although it has previously been claimed that chimpanzees afflicted with SIVcpz did not suffer the same debilitating side effects that HIV caused in humans, a 2009 paper by Brandon F. Keele at Hahn's lab found that SIVcpz-positive chimpanzees were 10 to 16 times more likely to die than chimpanzees without the disease. Since then, necropsy reports of chimpanzees that expired at Gombe have revealed AIDS-like wasting of the body.[381]

The decades of HIV/AIDS/SIVcpz research have failed to come to a concrete conclusion about the usefulness of comparing chimpanzee and human reactions to the viruses. However, once used in HIV studies, the chimps are then carriers of the virus and thus dangerous to human beings who handle them during the remainder of their lives. HIV-infected primates must forever be kept in isolation, or at least kept with only other HIV-infected individuals, in order to avoid disease spread. Ironically, it's even possible that HIV and AIDS entered the human population via early medical research and the injection of simian blood into human beings.[382]

Research on live animals is expensive and rife with the possible inconsistencies of individual reactions to various stimuli, not to mention that its results can be unreliable due to distinct reactions of various species. Although amendments to the Animal Welfare Act make provisions for the psychological well-being of primates in research facilities, animal welfarists believe that any primate living in a laboratory will suffer from enclosure and lack of mental stimulation, and this negatively affects their overall health and worth as research subjects.

During an 1989 conference entitled Well-Being of Nonhuman Primates in Research, Dr. Jeanne Altmann commented on "the inseparable nature of physical and psychological well-being."[383] Additionally, the diversity (independent of body size) throughout the primate kingdom virtually assures that even the most well-intentioned welfare stipulations will never be able to fulfill the various needs and preferences of all primates used in research.[384]

Luckily, as more and more major corporations are pressured by customers to phase out product testing on live animals (in fact, companies including GlaxoSmithKline, one of the major pharmaceutical corporations in the United States, have already forsworn chimpanzee research[385]), they are finding that alternative methods such as cell and tissue culture and computer models are more accurate and more affordable on the whole. Additionally, new products can continue to be introduced for sale if they are comprised of materials already proven to be safe for humans. As there have by now been decades of animal testing to determine the safety of various compounds, one would hope there will soon be a day when, without having to develop entirely new procedures and materials, manufacturers will rely almost completely on the plethora of ingredients that have passed tests and met safety standards and are already available for use.

Anytime the ethics of animal research are questioned, alarms about sentimentalism and a sort of anti-humanity malaise are raised. After all, science is historically objective, and, as Marc Bekoff explains, "animals are regarded as objects of study, not as subjects or experiencers of their own lives."[386] Without this distanced consideration of the world, scientific advances and the body of human knowledge would be quite different from what it is today. Primate research was used in the past and often continues to be used (albeit in smaller numbers) just because it's available and familiar. And yet, not discounting the important work done thus far, and while also not denying that information should continue to be sought, it appears, at least to some people, that using nonhuman primates in research is—for lack of a better term—antiquated.

An inquiry into the appropriate use of primates in modern research, *The Psychological Well-Being of Nonhuman Primates*, stated plainly, "Research procedures should not be sustained merely because they have been used in the past."[387] Change can be uncomfortable, certainly, and particularly in the realm of science, which is based on repetition and predictable outcomes. But as Justin Goodman of the Laboratory Investigations Department at PETA noted, in the past science has presented humans with disturbing and "inconvenient" facts that humans were forced to accept and learn to live with, and this adoption of new ideas cannot be avoided simply for the sake of convenience and comfort.[388] In another article, he added, "Self-reflection and scientific inquiry can lead to conclusions that are uncomfortable and inconvenient, but society will never progress if people choose to assimilate only the ideas that reinforce their personal biases and protect their own interests."[389] In other words, ignorance may be bliss, but it makes for some shoddy science.

In late 2011, a study was requested by the National Academy of Sciences and the Department of Health and Human Services of the National Institutes of Health with the purpose of analyzing the scientific necessity of chimpanzee use in medical research and without considering any related financial or ethical considerations. From the start of the analysis, the committee felt that "the chimpanzee's genetic proximity to humans and the resulting biological and behavioral characteristics not only make it a uniquely valuable species for certain types of research, but also demand a greater justification for conducting research using this animal model."[390] The committee concluded that although chimpanzee research has been useful in the past, "most current use of chimpanzees for biomedical research is unnecessary,"[391] except possibly for the completion of testing currently in process based on the chimpanzee model, specifically in relation to research on a Hepatitis C vaccine.

Recommendations included that the National Institutes of Health limit chimpanzee research except in cases where there is no other nonhuman model available for research, or when researching debilitating or life-threatening illnesses. In response to the committee's finding, the National Institutes of Health (the federal agency responsible for funding most research in the United States) announced in December 2011 that it was suspending all grants for new research using chimpanzees.[392] Although this landmark announcement pertained only to chimpanzees and not all primates used in research, many animal rightists viewed it as a stepping-stone to the eventual end of primate research as a whole. In addition, as of January 2013, the National Institutes of Health. has reported that it is considering permanently retiring most of its chimpanzees from use in research and moving them to sanctuaries.[393]

We have used nonhuman primates to learn about ourselves, and in the process we've learned how much we and our related species share. Yet clearly we are *not them*, and we continue to live in a state of dominance whereby we are able to do such things as conduct research on other species against their will. As Peter Singer states, "Indeed, the value of the great apes as research tools lies precisely in the combination of two conflicting factors: on the one hand, the fact that, both physically and psychologically, they very closely resemble our own species;

and on the other, the fact that they are denied the ethical and legal protection that we give to our own species."[394]

Is this so wrong? After all, is this not proof of survival of the fittest, a Darwinian sort of permission to use the power granted to us? Perhaps the most qualified people to answer questions related to the ethics of using primates in research would come from the field of primatology. Surely people who are experts on primate cognition and awareness must have some concrete evidence that could either support or decry their use in laboratories.

Primatologist Duane Rumbaugh emphatically states that "the more enlightened perspective is that these animals are related to us. The roots of what made us what we are can be found in them. They don't just appear to suffer...they do."[395] In 2010 alone, 71,317 nonhuman primates were used in US lab research.[396] That's a lot of suffering.

5. Threats

We have met the enemy and he is us. [397]

- Walt Kelly

Western culture, heavily European-influenced and continuously spreading to other parts of the world, has many inbred distinctions that subtly and blatantly reassert human superiority over animals. For instance, Christian-based religions, which traditionally have pervaded Western culture, teach that because humans were created in God's image, they are superior to animals. Animals are viewed as lowly and uncivilized, subject to their base instincts.

The distinct boundaries between humans and the other living creatures in a Christian world has been a point of contention when it comes to animal rights and animal welfare, particularly regarding the dividing line between humans and their closest nonhuman primate relatives. Perhaps the genetic proximity of apes to humans is one reason that humans have tried so hard to prove that nonhuman primates are separate, the "other," and not to be included in the moral code. To include a nonhuman within the Western moral sphere could potentially threaten the ethical world as many people know it, and people don't like change. They will work hard, and historically have worked hard, to find reasons to justify their habitually sub-par treatment of nonhuman primates.

The potential rights and freedoms of nonhuman primates all over the world have been trampled in many ways as their populations are affected negatively by changing conditions in their native environments. Their home ranges are often destroyed in a frantic search to make money from dwindling natural resources. Their social groups and families are decimated by hunters eager to maim, capture, and kill them to supply a growing taste for unusual meats, black market animal parts, and the illegal exotic pet trade. Often viewed as pests for simply attempting to

151

survive in areas that used to be their own but are now densely populated by displaced humans, nonhuman primates are frequently killed by local people who have not been educated about the endangered species living within their community walls.

One example of a fast-growing and soon-to-be dire situation that is harming primate populations involves orangutans of Borneo and Sumatra. The greatest current threat to orangutans is the proliferation of palm oil plantations built on land cleared of native rainforest. It's estimated that rainforest loss has caused the death of 3,000 orangutans in the past thirty years alone. Each year, 4.6 million acres of rainforest is destroyed in Indonesia, mostly due to palm planting, a highly lucrative pursuit.[398] Since this crop is so easy to grow, the wide availability of palm oil has led it to become ubiquitous in manufacturing, and it is used in increasing quantities in a wide variety of staple products such as toothpaste and margarine.

It's not just the orangutans that suffer due to palm oil utilization. The world's growing reliance on biofuel alternatives to petroleum products has encouraged the use of palm oil for that purpose. Although palm oil does not release as many greenhouse gases when burned as petroleum fuels, those same greenhouse gases are released into the atmosphere as land is cleared to create more room for palms.[399] As if that weren't bad enough, when orangutans wander onto the plantations surrounding their natural homes, they are often killed by humans who view them as pests or even as convenient meals.[400]

Shawn Thompson, author of *The Intimate Ape,* a book about his experiences among the orangutans of Borneo, expressed his frustrations and fears thusly: "The reality is inescapable. The facts are plain. The twelve-million-year-old species with the slow, leisurely pace and the long, quiet gaze is dying as a species. It is dying because of the way we dominate our planet. The way that our human population continues to grow and the way that we human beings devastate the rainforests, it will ultimately be the end of orangutans in the wild."[401]

The increased demand for palm oil and the resulting destruction of habitat creates smaller orangutan home ranges, which leads to smaller groups of orangutans who, with an average of one birth every eight years, are then unable to reproduce at a rate that sustains the group. Primatologist Herman Rijksen estimates that 93 percent of Borneo orangutans disappeared during the twentieth century[402]; although there were more than 300,000 orangutans living in Asia and China in 1900, now there are only 48,000 in Borneo and 6,500 in Sumatra.[403]

Due to public outcry against the major companies using palm oil, some have agreed to purchase only sustainably harvested palm oil or to use alternatives in lieu of the environmentally expensive lubricant. This is but a band-aid on a large open wound. Unless the world's demand for palm oil dwindles, the companies involved will always find ways to harvest it from areas such as Borneo and Sumatra, the very areas where orangutans also fight to survive. In 2007, President of Indonesia Susilo Bambang Yudhoyono signed into plan a decade-long effort to protect the remaining orangutans. Without help halting the logging, burning, and planting cycle, however, the president estimated that total extinction would await the world's orangutans by 2050.[404]

While human population growth and the subsequent clearing of land can be devastating to naturally occurring populations of apes such as chimpanzees and orangutans, gorilla populations in some areas have actually benefited from human interest under the circumstances. These peaceful, exotic primates are a highly sought-after visual treat for foreigners looking for an authentic African safari experience. The ecotourism business has been able to thrive alongside naturally occurring populations of mountain gorillas, and the resulting boon to local economies has meant that these primates are actually worth more when protected than when killed and sold for bush meat or other purposes.[405] Grauer's (eastern lowland) gorilla populations have also increased in certain areas. A survey in 2010 of native Grauer's gorilla populations in Kahuzi-Biega National Park in the

Democratic Republic of Congo revealed close to twenty more individuals living there than at the last count, in 2004.[406]

Gorillas can live surprisingly well in areas that have been selectively logged. Chimpanzees instead find smaller home ranges problematic due to increased interactions with neighboring tribes and the ensuing violence. Increased warfare can decimate chimpanzee populations, and orangutans don't fare much better.[407] Some optimistic research by primatologists Marc and Isabelle Ancrenaz has shown that orangutans in the Kinabatangan area of Borneo have been able to adapt to living in secondary forests (areas that had been logged but then re-grew), perhaps due to how rapidly changing and thus possibly stimulating it may be to the orangutans living within.[408] Although this silver lining does not in any way compensate for the clearing of natural forests, it is heartening to know that there is a middle ground where business can thrive and ecological interests can still remain protected. It would be best if the gorillas and all other forest inhabitants could live undisturbed, but it's a sad fact that the damage to their homes has been done, and what is needed now is to find the best way to pick up the pieces and allow human and nonhuman populations to coexist as peacefully as possible.

Hunting is similarly injurious to primate populations, as their slow reproductive rate means that it's more difficult for them to rebuild their families after an attack. Many primates reside in areas that tend to have high levels of poverty, and in a time when foodies all over the world demand exotic meat and are willing to shell out big bucks for it, disadvantaged people will go to great lengths to earn some money whenever the opportunity presents itself. The black market trade for primate meat has seriously affected African and Asian primate populations. It may even have seriously affected humanity: One theory of how AIDS originally afflicted humans postulates that it occurred when Africans ate primate bush meat infected with the simian version of the AIDS virus, an often naturally occurring and harmless virus in nonhuman primates.[409]

Ironically, the more difficult it is to procure primate meat, the more the prices rise, which only encourages the cycle to perpetuate itself. Soldiers in areas that are under martial rule may even be ordered to capture local wildlife for use as moneymakers with tourists and in the markets, as was the case in Zaire with the 1987 hunting of bonobos.[410]

The black market trade for primate meat and body parts helps fuel the hunt, but some of this practice is a cultural matter. In certain parts of the world, eating primate meat is promoted by local spiritual leaders and folk medicine devotees. Practitioners of African black magic, for example, have sought the ears, tongues, testicles, and fingers of silverback (dominant male) mountain gorillas to make potions for restoring human virility.[411]

Even when the exotic animal sought after by poachers is not a primate, the traps set up to catch other animals can often unintentionally harm primates, who can get limbs caught in the wire of traps and as a result sometimes die of gangrene or other infections. An investigation into the tactics employed by local Sierra Leonean hunters working to fill orders of the well-known exotic animal exporter Franz Sitter revealed tactics to capture wild chimpanzees that included dog chases, herding the animals into trees and then shooting the adults, and poisoning known food sources to capture the infants once the adults had perished.[412] Of course, not only chimpanzees are hunted in this way; almost every species of primate is under similar threats.

Although it will take a global effort to keep natural populations of primates stable and stop or reverse their continual decline, the United States remains a major importer of primates due to its relative lack of prohibitions against using primates in research, entertainment, and the pet and zoo trades. (Other countries have much stricter rules against importing primates, and more details can be found in chapter six of this book.) A study in 1989 illustrated what a pivotal role the United States plays in the purchasing of primates from wild land. In that year the US imported 26,479 monkeys. This number is especially significant when compared with the numbers of monkeys imported by other developed countries, such as 5,027 in the United Kingdom, 4,702 in Japan, 4,183 in the Netherlands, and

2,430 in France. It is apparent that the US remains a final destination for many imported monkeys.[413]

The U.S. Fish and Wildlife Service regularly releases the records on primate imports into the US, but even in this, one of the most developed and technologically advanced countries in the world, the import reports are incomplete and confusing. Records for the year 2010 neglected to include where the primates ended up, the percentage actually inspected by a federal official, and how many were dead on arrival.[414] Although any country's effort to curtail the importation of wild primates will make a positive difference, one can only imagine the impact if the United States, the country that imports the most primates, were to make some changes in the way it treats and regards nonhuman primates.

Infant primates are still sought as pets or as an entertaining attraction for a business, and they can draw a pretty penny: While a captive baby orangutan might cost only five dollars in Borneo, it's worth five hundred dollars in Jakarta, ten thousand dollars in Japan, and thirty thousand dollars in the United States.[415] Many of them die during transit to their destinations, due to dehydration, malnutrition, or injuries and shock sustained during their capture. Indonesian animal exporters estimate their mortality rate in transit as high as 32-71 percent,[416] and the International Primate Protection League (IPPL)'s tracking of a shipment of 2,000 African green monkeys to the US revealed that over one quarter of the animals died before reaching their laboratory destinations.[417] A 2013 report from the Great Apes Survival Partnership (GRASP) estimated numbers of illegally traded great apes from 2005 through 2011, which included (at a minimum) 643 chimpanzees, 48 bonobos, 98 gorillas, and 1,019 orangutans - and only 27 smuggling arrests were made in Africa and Asia in connection with these numbers.[418] Animal deaths often occur during the trapping/capturing procedure, possibly even ten deaths for every infant captured and sold at market.[419] GRASP claims that for every gorilla infant captured, fifteen other group members will have perished.[420]

Wildlife crime, long seen as "soft," is now virtually equal to the trade in drugs, weapons, and people in terms of its financial significance and the way it operates."[421] The extensive and dangerous networks of black market animal smugglers have a wide range of devices and methods to circumnavigate the legal barriers in the commercial exotics market, so more attention must be paid to this matter to deal with the system as a whole. Legal rulings that are not backed by traditional and/or cultural acceptance mean little in countries where human suffering is abundant and animal suffering becomes secondary, at best. Animal dealers and smugglers find ways to get around the current legislatures in order to make money doing what they do best, at the expense of thousands of animal lives. As primatologist Dian Fossey put so succinctly, "Active law enforcement is active conservation at work."[422]

Human poverty can also affect the well-being of indigenous primates negatively when it comes to land use. It can be difficult to justify conservation efforts to human beings who are themselves having trouble surviving, and land in many primates' home ranges has been claimed by local human communities for cultivation of crops or cattle grazing.

As human populations and demand for land increase, the availability of fertile land quickly decreases, along with the protected home ranges of many nonhuman primates. As Dian Fossey noted, "Foreigners cannot expect the average Rwandan living near the boundaries of the Parc des Volcans and raising pyrethrum for the equivalent of four cents a pound to look around at the towering volcanoes, consider their majestic beauty, and express concern about an endangered animal species living in those misted mountains. Much as a European might see a mirage when stranded in a desert, a Rwandan sees rows upon rows of potatoes, beans, peas, corn, and tobacco in place of the massive *Hagenia* trees. He justifiably resents being refused access to parkland for realization of his vision. American and European concepts of conservation, especially preservation of wildlife, are not relevant to African farmers already living above the carrying capacity of their land."[423]

Big business is yet another roadblock in the path of protecting the home ranges of nonhuman primates. A cycle of damage often begins when logging companies cut wide swathes of forests, clearing trees and building roads to make it easier for their machinery to cut down even more trees for wood. Such wide roads also facilitate hunting, making what was once deep forest more easily accessible to anyone with a weapon and a will to take anything provided by Mother Nature. Although land clearance may be appreciated by the local people, who can then grow more crops and fight widespread poverty, the land clearing mostly benefits the multinational corporations that want access to natural resources for palm oil and rubber.

Dian Fossey led gorilla research in Africa and was shocked when, in the late 1960s, a director in the local park service (an organization that normally fought against poaching activities) approached her because he wanted to obtain an infant gorilla to use for barter with a German zoo, which promised in return to provide the park with a much-needed vehicle and additional funding. Fossey realized that, despite her pleas and requests to respect gorillas' natural peacefulness and social ties, the official had larger concerns that required him to take what he could from the land in order to obtain needed supplies or political advantages. Although Fossey refused to grant his request for an infant gorilla, she later saw an infant he had obtained by working with one of the leading poachers in the area. Ten adults had perished in their fight to steal away the infant.[424]

Understandably, most people with interests in logging and poaching activities do not take kindly to environmentalists trying to stop their business. Violent threats, subtle terrorization by invoking voodoo that instills fear into the hearts of locals who understand the cultural significance, and retributions are not uncommon. Dian Fossey herself paid the ultimate price, giving up her life to defend the land of her beloved gorillas. Her murder remains unsolved to this day, but it is believed that she was attacked in retaliation by poachers who resented her intrusion into their business.[425]

Solutions to such problems often involve local governments, but in many areas where nonhuman primates live, the local governments can be unstable at best and war-torn and corrupt at worst. War violence affects not only the humans of the area; for instance, Grauer's gorilla populations dropped from 250 to 130 during a particularly violent war in the Democratic Republic of Congo.[426] In areas where the governments are concerned about ecology, some have made resolutions to halt the destruction of native species. For example, one of the proposed solutions to the hunting problem in poverty-stricken communities has been that governments should reimburse locals for the protein they lack when refraining from eating bush meat. Clearly this solution is tenuous and would be difficult to manage in areas where the governments most likely have problems they consider to be more dire. Increased tourism has also been proposed as a method of rejuvenating troubled African economies. Although this idea would have the potential to increase appreciation of native flora and fauna, the political instability of many African nations, as well as the possible commercialization that often accompanies rampant tourism, may result in further deforestation and depletion of natural resources.

In her book *Gorillas in the Mist*, Dian Fossey discussed the various kinds of conservation approaches available in the African mountains where her field studies of mountain gorillas took place. Active conservation involves direct contact with individual local inhabitants, such as people who know the land, language, and customs but may need supplies and certainly the funding necessary to fight to regain control of their native lands. Fossey felt that this had to be supplemented with a backbone of laws and policies, enforced by the government, to ensure that land is used properly and punishments are doled out for poaching and other illegal activities.[427]

Theoretical conservation is more abstract and long-term, although it need not exist independently of more active conservation methods. It may involve encouraging tourism to foster appreciation of the land, making improvements to roads and buildings to stimulate the local economy in legal ways, educating local communities to foster increased respect and appreciation for the species that live nearby, and even finding ways to habituate the native species to humans so that they may be viewed and perhaps photographed in their daily existence but not injured or threatened. Although she felt that the immediate threats to species such as her mountain gorillas must be solved first, Fossey believed that a combination of both active and theoretical conservation methods is best to realistically prepare native species and lands for the long-term cohabitation of human and nonhuman primates.[428]

Even if all the hypothetical conservation theories were completely functional and practical, there are other blocks in the road to picture-perfect conservationism. According to primatologist Craig Stanford of the University of Southern California, one is the myth of the noble savage; many cultures, American included, practically deify the rugged individualist who is able to hack out survival by relying on Mother Nature and the fruit of her womb.[429] There is a certain romance assigned to the person who can live off the land and become the king of the forest, rising triumphantly over the other species that inhabit the place.

It's unfortunate but worth noting that most of the areas where nonhuman primates naturally occur are in the midst of widespread poverty among the human populations. This means that humans are increasingly desperate to make a living any way they can; thus, survival by hunting or developing land so it's hospitable to the growing human population has a much higher priority than protecting the wildlife living within those areas. Primates that are able to subsist alongside local humans are often viewed as pests, as their natural intelligence leads them to take advantage of living near humans, whether through stealing their food or invading their homes. When local primates directly harm the very limited resources of humans, those humans are typically less likely to consider them valuable and worthy of protection.

Additionally, Stanford points out, although people assume that a better economy will lead to better protections for native lands and animals, this isn't always the case. Too often, increased funding improves the lives of humans living in the troubled area, and caring for the environment takes back seat. As endocrinologist Shigeo Honjo of Japan's National Institute of Health replied to IPPL's critique of his import of chimpanzees out of Sierra Leone and into medical research, "We, human beings, always want to be healthy and peaceful. Unless we are healthy and peaceful, we can not be considerate for the protection of chimpanzees. And the chimpanzee can contribute to the promotion of human health. This is the way of thinking of us, Japanese scientists. We usually call this style of thinking the circle of transmigration. Don't you agree with this thought?"[430] Cultural and personal beliefs affect the considerations different countries place on the world's natural resources and how humans should exploit them, if at all.

The extremely brief and basic lists in this chapter represent only a superficial skimming of the threats facing nonhuman primates. Most stem from humanity's insatiable need for growth and expansion at whatever the cost. Whether primates suffer in the wild due to destruction of their natural homes, being hunted down as food or labeled as pests, or being trapped and shipped off to laboratories or sent to live as surrogate children in human homes, it's all entirely interwoven, says Shirley McGreal, director of IPPL, explaining, "You can't separate out the research demands from the logging, from the pest-shooting, the pet trade. It's a cumulative effect."[431] As depressing as that may be, it's obvious that because humans have created the problems, only humans have the ability to fix this dire situation. Primatologist Geza Teleki believes, "Looking at chimpanzees from where I stand, eye to eye, not down my sharper human nose, I consider it sheer arrogance to perpetuate the anthropocentric view established by my ancestors simply because that was the collective human impulse."[432]

Perhaps one pathway to species conservation is through the heart. Author Shawn Thompson feels that the most success can be wrought by appealing to humanity's love for cute, intelligent animals. He wrote his book *Intimate Ape* in an effort to help document and advertise the plight of native orangutans in their dwindling home ranges, and he tried to make "people fall in love with orangutans, which then becomes the 'reason' for saving orangutans, the incentive. We need to feel that orangutans are part of our family. And of course, in terms of evolution and genetics they are."[433]

Thompson recognizes the danger wrought by the mass consciousness of modern people, who in many ways value economics, politics, and other personal concerns over the conservation needs of a species or the ecological requirements of the spaces in which they live. The damage caused to native primate communities has not occurred instantly, but over generations, and it will be in a similar fashion that improvements can be made. Educating people on the needs of the planet and her species, as well as the ramifications of living in complete ignorance of these facts, is necessary and is the first step toward spreading awareness about vanishing species such as the orangutan, chimpanzees, gorillas, gibbons, and other threatened primates living in hostile environments.

6. Legal Progress

It is time that judges consider that as the ancient foundations have begun to rot away, so the law of animals that rests upon them should be changed. [434]

- **Steven M. Wise**

Anti-cruelty statutes are perhaps the earliest forms of legislation that were prepared on behalf of animal welfare in the United States. Although today various states enforce anti-cruelty laws to varying degrees and with similarly varied punishments, all anti-cruelty laws rely on a specific owner/pet relationship. In order to encourage the proper treatment of an object that happens to be alive, regulations stipulate that owners must provide proper nourishment, water, shelter, medical treatment and humane transportation, and they prohibit abandonment, torture, poisoning, and unsanitary living conditions. [435]

Court cases have questioned humanity's direct duty to animals (inferring the animals' inherent desire to live free of pain, with sufficient nourishment, and so forth), as well as indirect duties and their effects, such as the commonly held belief that treating animals humanely is necessary to maintain public order. Exemptions to anti-cruelty statutes may be granted in the cases of animals involved in laboratory research, food production, or any other instance "justified when it is necessary to 'assist development or proper growth, fit the animal for ordinary use, or to fulfill the part for which by common consent it is designed,'"[436] such as training to complete certain tasks that would benefit humans. In instances when an owner has purposely taken the life of his or her animal, the courts generally do not see fit to interfere or prosecute, out of respect to the owner's discretion when dealing with his or her own property. In general, it is only when harm is suffered by an animal in an instance in which there is no financial gain at stake that an action is considered cruel or wanton.

163

United States law considers objects to be either things (objects of property) or persons (beings worthy of enjoying legal rights). A right, legally, is "recognized and enforced by the legal system,"[437] something that allows individuals to receive equal benefits, concerns, and considerations from others and protects them from manipulation for utilitarian purposes. Lawyers, professors, and animal rights experts consider a right to be a multifaceted composition of consideration of interests, ability to avoid evil, freedom of choice in daily decisions, and protection from the interference of others who may infringe on these liberties.

Current laws pertaining to animals exist to regulate their use by humans and ensure their humane treatment. All current laws have grown from past laws, so despite changes in public opinion that resulted in legal revisions and amendments, the laws that govern today's society are still largely based on the legal rulings of our ancestors. As lawyer Steven M. Wise described, "Raised by age to the status of self-evident truths, ancient legal rules mindlessly borrowed may perpetuate ancient injustices that may once have been unjust because we knew no better. But they may no longer reflect shared values and often constitute little more than evidence for the extraordinary respect that lawmakers have for the past."[438]

As lawyers rely on past legal cases to set precedents for current cases, rulings made during very different times are often reinforced indirectly in the courts today. In this regard the legal system currently in use in the United States, as well as in many European countries, still retains the influence of religion, in that man is considered superior to other living species. Although science is increasingly valued over religious texts in matters of governance, it takes time for laws to reflect this change in popular thinking.

Currently only human beings are deemed worthy of personhood in the US legal system, and although rulings can establish methods to treat primate "property" in increasingly humane and considerate ways, animal rights proponents ask that animals be granted personhood (something which sounds as though it must, by definition, involve being human, although this is not the case, and is a defense based purely on semantics). As a result, some laws use the term "animal guardian" instead of "pet owner" in order to avoid as much as possible the sometimes inflammatory consideration of animals as property; however, as of the writing of this book, animals remain legal property and recipients of very few liberties of their own.

Similar to the way that African American slaves were not granted true legal protections until the law recognized them as persons and not property, nonhuman primates must have their inherent rights and interests legally accepted before any comprehensive improvements in their treatment can occur. Until then, their lack of personhood means that they can only have legal standing in terms of being means to a humanocentric end.

Personhood need not be synonymous with humanness. Granting the personhood of animals does not require admittance that they are human, as, of course, by definition they're not. As Gary Francione puts it, "To say that a being is a person is merely to say that the being has morally significant interests, that the principle of equal consideration applies to that being, that the being is not a thing."[439] The benefits that accompany personhood, such as the right to liberty and the ability to perform natural behaviors, would be granted to nonhuman primates under this rights theory, but without personhood, future legal decisions must remain relegated to what Francione refers to as "micro-ethical issues."[440] Concerns such as cage size, enrichment requirements, and nutritional regulations only distract from the larger issue of granting inherent rights to nonhuman primates. In Francione's words, "Is our exploitation of nonhumans justified in the first place?"[441] Or have humans just grown ever more adept at crafting laws that make our exploitive treatment of nonhuman primates seem necessary and unavoidable?

Change comes about in small doses and typically only after tremendous efforts of individuals impassioned by a particular cause to educate those who make decisions. When a group of judges were presented with information about Project

Washoe (the group of chimpanzees who were taught sign language at Central Washington State University) in 1999, some of the judges' perceptions of personhood remained the same, but for others it was an eye-opening experience. Supreme Court Justice Faith Ireland summed up how she was affected, saying that it "was a paradigm-shifting experience for me and challenged some of my basic assumptions and presumptions on the subject.... The ethical challenges...have many parallels in our historic experience of judging, such as slavery, women's rights, and desegregation."[442]

Previous chapters of this book have discussed the difference between the movements for animal rights (which opposes any uses of animals) and animal welfare (which seeks to improve conditions under which animals are used by humans). Although primate activists are pleased with any legal advances that directly or indirectly benefit nonhuman primates, animal rights lawyer Gary Francione argues that the focus should be on animal rights, and he is against the creation of additional animal welfare laws. He believes that supporting the system of primate welfarism would only continue to permit the problems at hand for primates, such as living in unnatural, uncomfortable circumstances. Many of the animal welfare laws exempt the institutions that, coincidentally, use the most animals, such as research laboratories. Because animals are viewed as property, courts have traditionally upheld the belief that animal suffering is justified by human gain, either financially or intellectually, regardless of how insignificant or gratuitous that human gain might be.

Another reason that Francione opposes an increase in animal welfarism is that animal welfare cases usually involve criminal laws, and it can be quite difficult to prove that a defendant committed animal abuse with the "culpable state of mind" so inherent to criminal law because, in many cases, institutionalized practices openly support what many animal rightists and welfarists would consider to be abuse. Even in cases where malicious intent is proven, penalties are often minimal and not significant enough to act as a deterrent. Proper usage of animals is considered foremost, instead of what is actually in the best interest of the animals.[443] Although it is technically possible for animals to retain their legal status of property and still receive additional rights, that gain would fail to recognize an animal's inherent rights as a living being.

The United Nations' World Charter for Nature, adopted in 1982 to draw attention to humanity's ability to both quickly destroy the environment and also perform an about-face to halt further destruction, stated, "Every form of life is unique, warranting respect regardless of its worth to man, and, to accord other organisms such recognition, man must be guided by a moral code of action."[444] What's notable about this statement is that it considers very clearly the inherent value of organisms on their own and not only in relation to man.

There are more than fifty federal laws protecting animals in the US,[445] and although each state has its own anti-cruelty laws that apply to all mammals (and thus primates), in forty states the anti-cruelty laws do not apply to research facilities at all. In 2007 there were approximately 70,000 primates being used in research facilities (compared to approximately 15,000 either living as pets or working in the entertainment industry).[446] Thus, there is clearly a large community of nonhuman primates that are at risk for continued mistreatment unless humans intervene on their behalf.

In additional to professional accreditation groups and self-governed practices within the biomedical research field, research animals are protected at the federal level by government departments that include Agriculture, Health and Human Services (which includes the National Institutes of Health, Centers for Disease Control, and Food and Drug Administration), Defense, the Environmental Protection Agency, the National Science Foundation, and Veterans Affairs, with additional agencies linking departments whose areas of jurisdiction overlap.[447] Many of the subsequent laws the federal government has passed (as listed below) were responding to a very real need and are instrumental to providing a decent life to primates living in captivity.

The Lacey Act (1900)

Signed into law in 1900, the Lacey Act was one of the first US laws to defend conservation of species. It was created to prohibit trade in illegally captured, sold, and transported wildlife, fish, and plants.[448]

The Animal Welfare Act (1966)

The Animal Welfare Act (AWA) was federally adopted in 1966 as the Laboratory Animal Welfare Act. Renamed in 1970, it continued to be amended through the 1990s.[449] As discussed in the first chapter of this book, the AWA was integral to establishing basic standards of care for warm-blooded, vertebrate mammals living in, or destined to live in, captive facilities in the United States, including those being subjected to research (but excluding retail pet stores and pet owners). The law authorized the Animal and Plant Health Inspection Service (APHIS) of the United States Department of Agriculture to demand provenance records, license, and inspect facilities holding animals for certain reasons, such as those that tested the animals or exhibited them as part of a show. Violators of the AWA faced penalties and revocation of licensure.

The goal of the AWA was to halt the illegal black market trade in animals and to ensure that the legal system governing animal commerce resulted in animals living in humane conditions. Research facilities were required to create Institutional Animal Care and Use Committees (IACUCs), with at least one trained veterinarian, to inspect and observe animals' living conditions biannually and to maintain a basic level of care for research subjects. With the inclusion of later amendments to the law, minimum standards of care were set to insure that animals received humane treatment (including anesthesia) during experimentation. [450]

The most important amendment to the AWA occurred in 1985 with the publication of "U.S. Government Principles for the Utilization and Care of Vertebrate Animals Used in Testing, Research, and Training." This document listed additional care specifications for nonhuman primates in such topics as transportation, sanitation, and mental health. These principles advocate the consideration of alternative research models, such as computerized or mathematical models, as well as advising researchers to imagine that any activities that may cause pain in humans should be assumed to cause similar pain in research animals. The ability to form natural social groups whenever possible, access to environmental enrichment, the official scientific consideration of aggressive tendencies exhibited, and species compatibility were all listed as topics worthy of scientific concern and investment, so as to best suit the individual member of the species being studied. The addition of elements pertaining to psychological well-being of a nonhuman primate is considered the first regulation of its kind in American federal law.[451] The 1985 amendment also marked the first time that Congress intervened in the activities and processes of scientific research.

The fact that the AWA is a federal law raises a question: Why was the federal government concerning itself with animal welfare? The United States Constitution does not mention animals at all. Historically, the state legislatures discussed animal welfare, especially when it came to liability in cases of animal abuse or harm. However, complexities arose with interstate transport of animals. An umbrella of organized protection was needed to oversee animal handling within the United States, and thus the AWA was born.

It's crucial to realize that the AWA acts under the assumption that animal testing is ethically permissible. It was created not in response to an animal rights debate, but as a resolution to ensure animal welfare in applications of animal use. The AWA controls husbandry issues, not the purposes of or methods used in biological research. Monitoring the accurate application of anesthesia is the only interference USDA inspectors are authorized to perform by the AWA, other than confirming that researchers considered using non-animal models and that the experiment is not a duplicate of a prior experiment.

According to Francione, "Animals are treated as the property or 'resources' of the facility, and the only concern of the

AWA is to ensure that these resources are used efficiently, which, in this situation, means that they produce reliable scientific data.... To the extent that the AWA recognized animal interests, those interests may be sacrificed as long as there is some benefit involved."[452] Although the AWA stipulates that each facility have an IACUC in order to officially maintain humane living conditions, the IACUCs are comprised of animal researchers, who not only may be unlikely to find fault with vivisection overall, but who also may be reluctant to point out flaws in the systems of their colleagues.

Other problems raised by the AWA's specifications involve semantics. Although the AWA was created to minimize animal suffering, how exactly can an animal's distress be measured? The mammals covered under the AWA vary widely in their behavioral tendencies. Considering that a chimpanzee will smile in fear, looking extraordinarily like a very happy human, there are likely to be many reaction differences in species that are even less closely related to humans biologically.

Despite conferences in 1987 and 1989 entitled "Well-Being of Nonhuman Primates in Research" and involving primate experts and scientists joining together and seeking answers on how best to infer a primate's comfort level in laboratory environments, the USDA was not able to come up with a definitive methodology of pain detection. Where is the dividing line between discomfort and distress? Factors contributing to an animal's well-being were considered to include health, production and reproduction, physiological or biochemical evidence of reactions to stress, and behavior,[453] although certainly a research animal could appear physically healthy but be suffering mentally. Even if someone were to establish an answer to such a subjective question, the next issue to arise is that one individual's minimal distress may well be another's maximum.

The AWA prohibits inflicting "unnecessary" pain in animals, which is ambiguous and therefore troublesome; for if researchers decide that pain is indeed necessary in a particular experiment, the AWA grants them license to continue. Richard

Brown, formerly the senior program veterinarian at the University of Wisconsin at Madison, was frustrated that "[researchers] don't have to prove that [the use of painkillers] will interfere.... They only have to say that it *might* interfere."[454] Pressures to avoid making waves, or to keep quiet to keep one's job, combined with bare-minimum reporting to the appropriate agencies, wind up causing many others who may potentially be sympathetic to the research subjects to keep their mouths shut.

Efficiency is key, as the laws function as a guarantee that animals are best used for the purpose for which they were intended[455] (which seems to contradict having the animals' best interests at heart). For example, most primate species are highly social, yet the standards of scientific protocol of biological research often prohibit communal living spaces for the research subjects so as not to transmit pathogens or cross-contaminate subjects. In instances when an experiment is designed in opposition to AWA husbandry standards, such as an experiment that causes seemingly unjustified pain, researchers can simply write up a report to their facility's IACUC, admitting awareness of the pain inherent in their experiment, and avoid further investigation or review.[456] The IACUC only has authority to disband an experiment if an animal's pain cannot be justified at all.

The AWA holds researchers liable to "consider alternatives to any procedure likely to produce pain to or distress in an experimental animal."[457] In a scenario following those guidelines it would be acceptable if an animal were writhing in pain, because this wasn't the "likely" outcome hypothesized by the researcher. Requiring a researcher to merely "consider" something, instead of actually doing something, is tantamount to grasping at the grains of sand of animal welfare. As tightly as the granules may have been held, they escaped and were lost, and animal suffering was not only taking place but was legally sanctioned.

USDA inspections to uphold the AWA can be infrequent and not thorough. This is partially due to logistics. A 1994 study showed that the western division of the USDA was responsible for 13 states' worth of facilities. The approximately 3,000 investigations due each year were split among only18 veterinarians.[458]

Adding a component ensuring mental health of research subjects not only added more work to an already difficult task, it added a murky layer of subjectivism, as well. After all, the National Institutes of Health has declared that "there is absolutely no indication that the peer review process has in the past, or is currently, serving as an effective forum for ethical review."[459] Although regulations do stipulate that the Animal Care Committees required of each facility include at least one member who is not an employee or in the immediate family of an employee of that facility, animal rights groups have fought to require at least one public member as a part of the committees. This would potentially provide some transparency and impartiality in the group.

An investigation by the USDA in 2005 found that many research facilities were not researching alternative methodology, searching for duplicative projects, or providing adequate veterinary care.[460] Because 33 of the 50 worst violators were educational institutions, The Humane Society of the United States (HSUS) asked schools to develop internal programs to prohibit suffering and to more carefully monitor their living research subjects. As of 2011, more than 60 universities had committed to the cause.[461] It is estimated that approximately 600 educational facilities use some sort of animal research, and an annual $11.5 billion dollars of federal funding went towards research at 420 of those schools.[462] When the HSUS reviewed incident reports during a three-month period, they found that more than 800 animals had suffered needlessly, due either to poor experiment design, lack of foresight, or sloppy maintenance and handling within the laboratory.[463]

A test of the AWA arose the mid-1990s with the case of ALDF v. Glickman, when a man named Marc Jurnove observed the primate display at the Long Island Game Farm Park and Zoo. He was displeased with what he considered to be a general lack of consideration for the primates on display, including a chimpanzee housed alone and squirrel monkeys agitated by being housed within viewing distance of a bear. After he contacted several agencies, including the USDA, to report the lapses in care, investigations were carried out, and it was decided that the park was indeed upholding the legal standards for animals in its care.

After further back-and-forth complaints and investigations, Jurnove grew frustrated with the apparently legal compliance of the park, which seemed in direct opposition to the uncomfortable living conditions of the park's primates. Frustrated by his seeming inability to help them escape such conditions, Jurnove eventually sued on behalf of the highly intelligent animals he saw distressed and suffering emotionally in unhealthful and stressful living conditions. Prior judicial decisions had upheld an individual's right to enjoy displays and views without being visually assaulted by upsetting situations, and the law in this case was more concerned with a human's interest in viewing animals free from inhumane treatment than the well-being of the animals living destitute lives. In ALDF v. Glickman, the judge granted that the aesthetic violations witnessed by Jurnove were sufficient grounds to advocate for improvements in the psychological well-being of primates.[464]

It's difficult to tell whether this sudden consideration was due to a true recognition and appreciation for primate intelligence, or if it was simply to smooth the path so that future zoo and lab keepers could continue to keep nonhuman primates in their care. Even with this change, the AWA has much room for improvement. Criticisms of this law include its failure to include certain topics that can greatly affect animal life, such as the monitoring of pet shops and of animals used in pre-college education. Additionally, it can often be difficult to tell if a nonverbal animal is truly psychologically healthy (especially if the animal has been living amidst laboratory testing that may affect physical reactions to the environment), and thus it could be argued that this law does not go far enough toward protecting such a cognitively evolved species. Still, it would be difficult to disagree that captive-held primates benefited from this development. In addition, the amendments and fine-tuning of this body of rules accumulate over the years, acting as evidence of the public's growing consideration of animal welfare even in the face of its increasing reliance on science and research.

The Endangered Species Act (1973)

After the unquestioned post-war consumerism of the 1950s and 1960s in the United States, Americans once again turned their focus on the earth and her natural resources. The Endangered Species Act of 1973 (ESA) developed a classification system of species that were threatened with extinction, in the hope that attention would save these animals from disappearing altogether. New protections and actions to conserve natural ranges were meant to ensure that future generations of troops, groups, flocks, and packs would continue to roam the earth. A species was classified as "endangered" if it was near extinction, granting it strong protections from use by humans. A species was listed as "threatened" if it was not yet endangered but would be headed along that path without further human intervention; thus, the rules regarding the handling of threatened species were less strict than for endangered species.

Nonhuman primates are included in these species classifications, but not always in what might seem predictable categories. For example, chimpanzees have a strange split listing; they are endangered in the wild but only threatened in captivity. A special exception in the ESA excluded from consideration any animal living in captivity at the time of the law's passing (in 1973), including their offspring, as well as any animal of that species born into captivity after 1973.[465] Essentially, this law applies only to primates living in the wild and those that have been plucked from their natural homes after 1973 and brought into captivity. Countries in which chimpanzees may have their home ranges are listed, and any chimpanzees originating from these ranges are automatically included under the ESA's protections (unless they are exported illegally, in which case the ESA cannot help them).

It's known that a captive chimp born in Washington State and a wild-born Tanzanian chimp are biologically similar. They are not two different species; yet, one is offered more protections by the ESA. There is a large community of primates, including chimpanzees, being bred in the United States. Those hundreds of thousands of primates bred to supply the nation's labs, zoos, animal trainers, and exotic pet stores are not considered endangered, even though, were these animals able to survive in their natural habitat (a highly unlikely event to begin with), there wouldn't be sufficient natural resources or even natural space to accommodate them in a reasonable way. Ironically, the high numbers of primates living solitary, unnatural lives in captivity has helped to move them off the endangered list and away from some potentially helpful protections.

Because the ESA regulates only the commerce of animals, it does not make considerations for gifting or trading of animals. Therefore, a primate breeder could claim to have given someone a capuchin, let's say, and there would be no legal ramifications under the ESA to consider, even if the capuchin was otherwise considered threatened or endangered. It was born in captivity and, thus, not protected by the ESA. According to author Alan Green, some "dealers with captive-bred wildlife registration permits, which authorize the sale and purchase of nonnative listed species that have been born in the United States, know that they can freely mislabel and otherwise lie about their transactions, and they do so with a vengeance.... And so the dealers and roadside-zoo operators, who need only fill out a relatively simple application to get a captive-bred wildlife registration permit, write 'donation' or 'gratis' on their annual reports and continue to commit their crimes with impunity."[466] Ownership of endangered and threatened animals can be transferred over and over again, despite a lack of proof that the individual owners are capable of properly caring for the animals in question.

In late 2011, in response to pressure from such primate support groups as The Humane Society, The Jane Goodall Institute, and the New England Anti-Vivisection Society, among others, the U.S. Fish and Wildlife Service announced a review of the chimpanzee's split threatened and endangered status under the ESA.[467] Although there have been other times in the past when the chimpanzee's status was reviewed, and captive chimpanzees remained listed as threatened instead of endangered, at the time of this writing, the ESA's listing of chimpanzees is still under review, and the final decision in response to the newest petition remains unknown.

The Convention on International Trade in Endangered Species of Wild Flora and Fauna (CITES) (1975)

The Convention on International Trade in Endangered Species of Wild Flora and Fauna, commonly referred to as CITES, is a treaty agreed upon by 175 countries (as of the writing of this book) that addresses the protection of species from endangerment and/or threat as a result of commerce. It was officially signed into practice on July 1, 1975. CITES governs the protection of approximately 5,000 species of animals and 28,000 species of plants through a system of appendices that rank the level of threat posed to each species' perpetuation.[468]

Every primate species is listed for consideration by CITES. Appendix I, which lists the species most threatened with extinction from hunting and other trades, includes gorillas, chimpanzees, and many other primates such as various lemurs, marmosets, gibbons, and monkeys. CITES dictates that all Appendix I species are protected from commercialized trade between member nations, or any trade at all if, due to trade, a species' survival in the wild would be at risk.[469] All remaining primate species are listed in CITES Appendix II, which means that they could be threatened with extinction if their trade is not monitored.[470]

CITES-participating countries are required to take steps to develop programs protecting their indigenous species from commercialized trade and misuse. The organization is a sort of self-governing body, with no strict disciplinary measures in the case of noncompliance. Political advantage can motivate the reporting of scientific data but also encourage selective ignorance. Should a member country wish to engage in outright commercialized trade prohibited by their CITES agreement, they can write exemptions into their local laws to justify the practice.[471] For example, the United States relies upon the stipulations of the Endangered Species Act as support for CITES. There have been times when primates bound for research were exempted from CITES regulations due to the trend-driven worthiness of the research and the perceived immediacy of further testing, such as during AIDS research in the 1980s and 90s, and also due to the subjectivity of an acceptable level of commercial success of any given research project that requires CITES-affected animals.

CITES is clearly not foolproof, and it has been criticized as being too lofty to affect dealings at the level of reality. Oftentimes CITES regulations can conveniently be overlooked when a member country wishes to barter with highly in-demand goods in exchange for specific primates. The years of education that are necessary to make true changes are expensive. Countries who agree to take part in CITES often don't have the oversight or funds to combat beliefs and practices that are deeply ingrained in local culture.

For instance, bush-meat trade in African countries is still rampant because human populations are expanding, poverty is increasing, and, quite simply, people will manipulate the resources available to them in order to survive. Local greed is more common than ethical dealings with regulatory bodies. Cultural norms are more powerful than governmental rulings that are not enforced or are poorly managed. As Mark Jones of Care for the Wild International, a UK-based conservation charity, bluntly analyzed the situation in a BBC opinion piece, "Exploitation of, and trade in, wildlife and wildlife products is driven by demand. In an ideal world, we would control trade in endangered species by reducing the demand, by educating people in consumer states. However, in the face of criticism concerning 'interference with national sovereign rights,' 'cultural traditions,' and 'ignorance of poverty,' such efforts are unlikely to succeed— certainly not in time to save many of the species...."[472]

Proposals to improve CITES vary, from increased funding to tougher restrictions to the organization somehow working more independently from the tentacles of political interests—aiming to work with the interests of endangered flora and fauna held as being of the utmost importance.

An interesting side effect of CITES is that the primates currently living in captivity in the US may be the last living bridge between the transplanted species and their native past. No primates have been imported into the US legally since CITES' adoption in 1975. As some species have particularly long life spans (chimpanzees, for instance, may live to be fifty years old), there are still primates living in captivity that were born in the wilds of other countries. That said, there is already a large population of primates bred in the US that, never having lived in the wild, have developed certain mannerisms, skills, and preferences suited to a life among humans. Never having learned survival skills from their parents, such as how to prepare a nest in a bush, how to extract gum sap from a tree, or how to hunt the next meal, these primates are quite adept at using utensils and opening bottles and are more comfortable on concrete than on grass.

Steve Ross, a behavioral scientist and captive chimpanzee specialist at Chicago's Lincoln Park Zoo, has noticed that "we're at a really interesting time in American history, when it comes to the state of chimpanzees. Probably in the next fifteen to twenty years, we'll get down to the last wild-born chimpanzee. At that point, the entire population will have been born here in the United States."[473]

It's possible that at some point the captive-bred species members may evolve so much that they will be considered a species separate from their relatives who were able to continue living naturally in the wild. From the capuchins bred to help the disabled to the macaques bred as pets, the gorillas bred to stock zoos, or the chimpanzees bred to supply biomedical testing, these species may slowly be changing in essential ways because of CITES.

The Health Research Extension Act of 1985 (1985), part of the U.S. Public Health Service Act

The Public Health Service Act is a US federal law enacted in 1944 that has grown to cover many diverse aspects of public health that the National Institutes of Health (NIH) oversees, including health insurance, cancer research, family planning, drug use, and mental health. In 1985 it was amended with the introduction of the Health Research Extension Act, with the goals of ensuring proper care for animals used in federally funded research, including responsible use of pain relief and veterinary care, when needed. Animal care committees were required to maintain accurate records at each research site in order to ensure compliance. If annual reviews by the NIH revealed that a facility was not in keeping with the standards promoted under the Health Research Extension Act, federal funding and grants could be withdrawn.[474]

This act permitted the federal government to continue funding biomedical research while also assuaging concerns regarding animals used in said testing. Federal funding is highly important to much research and is vital for the continuation of progress in the medical sciences; thus, researchers have a great incentive to follow the guidelines and better care for the animals in their facilities.

The Depictions of Animal Cruelty Act (1999)

Although this ruling did not arise specifically in relation to primates, it can affect primates and their proper handling. The Depictions of Animal Cruelty Act was signed into law in response to media depicting the intentional torture or harming of animals. Specifically, it was created to combat the sudden visibility of online videos or images of pets, particularly infant animals, showing them being harmed and crying out in pain. Any person found to have created or distributed images depicting animal cruelty is subject to fines and imprisonment of up to five years.[475]

Two of the main problems that prompted this law's creation were the underground dog-fighting market and what are known as "crush" films (videos of animals being stepped on or purposely injured, often displayed on sexual-fetish websites). However, in April 2010, the law was overturned in US Supreme Court on the grounds that it was too broad, in violation of the free speech guaranteed by the First Amendment of the Constitution, and that it could be applied to industries such as hunting, for which the law was not originally intended. A renovated and more specific version of the 1999 act, renamed H.R. 5566: Animal Crush Video Prohibition Act of 2010 was signed into law in December, 2010.[476]

The Chimpanzee Health Improvement, Maintenance, and Protection (CHIMP) Act (2000)

Once it was clear that the protections for primates would continue to grow in both width and depth, it became necessary also to expand the stipulations of how certain primates should be treated when living in captivity. Additionally, specific funding was needed to ensure the proper care of primates throughout their lifespan. When chimpanzees become adolescents, for instance, their growing strength and mental cunning means they are no longer easily trainable for the entertainment industry and are often then sold to research labs. Even if they did not start life in a lab, many chimps end up there, and their long lifespan means that research subjects very often outlive the research for which they were bred or captured and assigned.

The CHIMP Act detailed the proper care for chimpanzees that had been retired from research conducted or approved by US governmental agencies such as the National Institutes of Health or the Food and Drug Administration. Under the CHIMP Act, chimpanzees could no longer be euthanized when not needed in research, except when humanely required. This was the first US law to restrict a lab's authority to euthanize.[477] Federal funds (90 percent of startup costs and 75 percent of maintenance costs) were provided to sanctuaries so that retired chimpanzees could have lifetime care.[478] Although they are often referred to as "surplus" (a term that seems more appropriate to describe unneeded beakers or textbooks), this was the first law to look past the years spent in research and recognize that chimpanzees deserved to have a certain quality of life for the remainder of their lives.

The CHIMP Act leaves the decision of when to retire chimpanzees up to the individual labs, which has the potential to involve conflicts of interest. The sanctuary system was deemed the proper location to place surplus chimpanzees, once retired, as well as any that are surrendered by their rightful owners. Of course, in accordance with the Animal Welfare Act, any approved sanctuary must have adequate provisions for proper housing of the chimps, so as to ensuring an individual's well-being. Chimps covered by this act may not pose a threat to other chimpanzees or humans, they must not be euthanized except when deemed medically necessary, and they may not be discharged from the protection of the CHIMP Act.[479]

Data on such chimps may be collected during noninvasive, routine examinations, but only if they cause minimal physical and mental stress (to the individual as well as his or her social group). Fees charged upon a chimp's entry into the system are used to maintain the system and its operation. A chimp may be denied access to the sanctuary system if there is not a location adequately staffed or prepared to accept and care for it. Additionally, if the complete medical, research, and ownership provenance record is not available, a chimp is not allowed access to the sanctuary system.[480] This roadblock was created in order to discourage black market exotic animal trade.

Stipulations are in place for the nonprofit sanctuaries regulated by the CHIMP Act, as well. There must be in place a governing board of voluntary advisors, replete with experts in the fields of primatology and zoology, as well as those experienced in business management, laboratory research, and biohazard handling. Proper reporting procedures must be evident.[481]

In 2002, Chimp Haven of Keithville, Louisiana, was chosen as the sanctuary to which primates retiring from experiments would be placed.[482] The board of Chimp Haven (comprised of several individuals with histories of experimenting on nonhuman primates) would have discretion to decide if any residents would be permitted back into research, depending on the merits of the proposal, with the Secretary of Health and Human Services having the final say.[483]

Although the CHIMP Act states that "no one who has been previously 'fined for, or signed a consent decree for, any violation of the Animal Welfare Act' can use a sanctuary chimpanzee for any research,"[484] a loophole in the CHIMP Act did not prohibit a chimp from being resold to a lab for further tests if this would cause minimal stress to the subject and his or her social group, if the research was dire or unique, or if deemed important (the scale of such determinations is unknown). In response to this loophole and due to public outcry, the "Chimp Haven is Home Act" amendment was signed into effect in December of 2007, ending the possibility of a return to research for the nation's retired chimpanzees and ensuring them an additional degree of safety for the rest of their lives.[485]

Additional International Laws Regarding Primates

As in the United States, the legal protections and status of nonhuman primates in other countries change continually, and even a carefully researched list reflecting the current state of primate legal affairs throughout the world would surely be obsolete by the time this book appears in print. The following information reflects interesting international developments occurring at the time of this writing.

When a positive shift in the ways that people consider the welfare of nonhuman beings occurs in any country, it has the potential to spread to other areas and cultures. Although testing on nonhuman primates seems necessary and integral to the medical and scientific community in the United States, it's promising to see that other countries are managing to make progress without relying on this system.

In 1997, amid discussions of other hot-button topics such as international human trafficking and terrorism, members of the European Union signed the Treaty of Amsterdam with the "common aim of improving respect for the welfare of animals as 'sentient beings.'"[486] The treaty officially went into effect in

1999, and it avowed that member countries were committed to keeping in mind animals' capabilities to feel pain and fear when under duress, as well as happiness when well treated, when considering matters of animal welfare.[487] Also in 1997, the United Kingdom stopped approvals of new proposals for research on certain primate species and has since developed a system of determining alternatives so that primate research is no longer needed.[488]

A meeting in September of 2005 between representatives of more than twenty countries in Kinshasa in the Democratic Republic of Congo resulted in the signing of the Kinshasa Declaration on Great Apes. This document detailed a commitment to combat the threats facing great apes and emphasized interdepartmental and international cooperation in order to develop National Great Ape Survival Plans, which required, among other things, that long-term, ecologically sustainable alternatives be made available to local communities so that home ranges of wild primates be less threatened. The Kinshasa Declaration on Great Apes, signed by representatives from Angola, Burundi, Cameroon, Côte D'Ivoire, Ghana, Guinée-Bissau, Indonesia, Mali, Nigeria, Republique Centrafricaine, Democratic Republic of Congo, Guinea, Senegal, Uganda, France, Tanzania, Italy, Belgium, Japan, Sweden, The United Kingdom, and The United States, set the lofty goal of a secure future for all great apes by the year 2015.[489]

A huge advance occurred in 2005 when a Brazilian court decided that a chimpanzee named Switzerland (a specific individual chimpanzee living in a Brazilian zoo with extremely crowded conditions) might have the legal personhood to obtain a writ of habeas corpus, meaning that the chimpanzee had the right to be legally represented by a human advocate in a court of law so that the legality of the chimpanzee's confinement could be examined. Brazil's constitution states, "The right to habeas corpus is granted whenever someone suffers or feels threatened by violence or coercion in their freedom of movement by illegality or abuse of power."[490]

Heron Santana, a Brazilian prosecutor, joined forces with animal rightists, students, and professors on behalf of 23-year-old Switzerland. Sadly, Switzerland was found dead the morning after her case was won, so she was never able to enjoy the freedom that had been granted to her. Nevertheless, the fact that a status of legal personhood was considered for a nonhuman, albeit for a short time, means that it could one day happen again (and perhaps for a greater population of nonhumans than just one individual suffering an unfortunate life).[491]

The Great Ape Project, started in 1993 by Peter Singer and Paola Cavalieri, celebrated a landmark victory in 2008 when the Spanish parliament voted in favor of primate rights. In move that many hoped would pave the way for similar decisions in other countries throughout the world, Spain dedicated itself to the Declaration on Great Apes, guaranteeing life, liberty, and protection from torture to apes (a year before that, the Spanish autonomous Balearic Islands had also endorsed the Declaration on Great Apes).

The Christian Science Monitor reported, "According to the declaration, apes may not be killed except under 'strictly defined circumstances' such as self defense. They may not be imprisoned without due legal process, and they may not be subjected to the 'deliberate infliction of severe pain,' even if doing so is said to benefit others."[492] It became illegal in Spain to conduct harmful experiments on apes or to keep them captive for entertainment purposes. Keeping apes in zoos would not be prohibited, but animal-care practices in zoos would be scrutinized, if not changed for the better. At the time of the legal victory, Pedro Pozas, Spanish director of the Great Ape Project, declared, "This is a historic moment in the struggle for animal rights. It will doubtless be remembered as a key moment in the defense of our evolutionary comrades."[493] Indeed, passage of the bill made Spain the first country in the world to fully recognize the rights of great apes.

The future of primates in research has steadily been brightening within the last few decades in various countries around the world. Since 1999, New Zealand has prohibited the use of nonhuman hominids (gorillas, chimpanzees, bonobos, and orangutans) in laboratory testing, except when it is *in the best interest of the nonhuman hominid* itself. This is practically the opposite of the situation in the United States, where testing is permissible precisely because it benefits humans and not the research subjects themselves. In 2006 Japan banned chimpanzee research when considered invasive, whether medical or pharmaceutical in nature—anything that would significantly affect their natural behavior as seen in the wild. By 2008, research on great apes had also been banned in the Netherlands, Sweden, Australia, Belgium, and Austria (where research on gibbons is also prohibited).[494] In 2010, the European Union banned research on great apes and tightened regulations for other primates, except research required for species conservation or if there was a "serious pandemic affecting the human population of Europe."[495]

In 2010 Brazilian courts revisited the case of a chimpanzee's possible denial of rights on grounds of habeas corpus. Great Ape Protection, on behalf of Jimmy the chimpanzee, argued that the quality of his conditions in a zoo outside Rio de Janeiro meant that he was being denied his freedom of movement and his ability to lead a normal life. As of this writing, the case is not yet resolved, but Great Ape Protection is determined that Switzerland's death will not be in vain, in hopes that perhaps Jimmy will be released from the zoo where he "is treated as a slave to produce economic benefits to humans, who do not care for his basic rights, which should be respected."[496]

The past few years have seen the proposal and, in some cases, the failure of new primate protection laws. One of these was the Great Ape Protection Act of 2009 (H.R. 1326), which would have prohibited invasive research on great apes, as well as breeding, purchasing, transporting, or in any other way using great apes for the purpose of research. It defined invasive research as anything causing physical or psychological pain, distress, or fear, up to and including death, and required the federal government to provide permanent retirement to great apes undergoing such invasive research at the time of the law's proposed passing. Although this bill was proposed to the US Congress in March of 2009 and again to the Senate in 2010, it did not pass and was never signed into law.[497]

In April of 2011, The Humane Society of the United States (HSUS) and the Physicians Committee for Responsible Medicine joined in Washington, DC, with Congressman Roscoe Bartlett of Maryland to propose the Great Ape Protection and Cost Savings Act (H.R. 1513/S. 810), which aimed to provide taxpayer savings while also permanently retiring great apes from invasive research. Representative Bartlett, who has a doctorate in human physiology, publicized his belief that the constantly evolving fields of science no longer required the use of great apes in research. An estimate by HSUS claimed that 80 to 90 percent of laboratory chimpanzees are currently not even used in research, and that no longer using great apes in research could save American taxpayers $25-$30 million dollars a year.[498] Unfortunately the Act was not passed during the 112th legislative session of Congress.[499] In January of 2013, however, the National Institutes of Health moved to retire to sanctuaries most of their 451 chimps used in research[500].

The recent spate of primate-friendly legislation introduced both in the United States and elsewhere in the world is encouraging for the future of nonhuman primates living in the wild and in captivity. Even in the case of proposed laws that fail to be adopted, whenever issues affecting the quality of life for primates are discussed, the conversation legitimizes and raises awareness of ways to improve their legal protections.

7. Primate Protectors

To give the orangutans a choice is very important for me.[501]

– Biruté Galdikas

Throughout history, it's proved obvious that in both the human and the animal kingdoms, "the less able a group is to stand up and organize against oppression, the more easily it is oppressed."[502] Just as it was necessary for groups of the majority to band together to protect a troubled minority in human history (whether in cases of civil rights, anti-Semitism, or the feminist movement, for example), human populations have joined forces to fight against what they perceive as threats against the well-being of nonhuman primates.

It wasn't until the 1960s that primatologists began to band together under the umbrellas of organized international conferences and journals. Once this collective aim was achieved, the great questions relevant to the field could be posed and intelligently answered by those who knew best. Field methodology and accuracy in reporting even the most mundane of primate behaviors were discussed and expounded upon so that primatologists the world over could assume some common ground and share practices. Although some earlier primatologists still clung to the belief that interference with the animals was required and carried no negative side effects, their opinions died out over a few decades, and today's observational, more silent, and less intrusive methods prevailed. Most important, the more modern protocol dictated that researchers must avoid all interruptions in the natural lives of their primate subjects because this would not only harm the individuals and their group dynamics but would also skew research results. The goal was to observe and record the natural habits and behaviors of these

187

fascinating creatures that were so quickly being affected by growing human encroachment.

International Primate Protection League

The International Primate Protection League (IPPL) was founded in 1973 by Dr. Shirley McGreal in response to the abundant profiteering trade in gibbons that she experienced firsthand while living in Thailand.[503] IPPL focuses on issues of conservation and ending animal research in order to protect the rapidly shrinking populations and natural homes of primates living in the wild. The projects and accomplishments of IPPL read like a laundry list of hoped-for legal precedents and cultural awakenings, from the 1970s through the present day.

In the early 1970s, work spearheaded by McGreal, illustrating the fates of military research primates in the United States, resulted in a ban in India and Bangladesh on exporting macaques.[504] An expose published by IPPL in 1974 revealed a vast network of primate smugglers, which was thereafter shut down. Project Bangkok Airport was carried out the following year. In this effort, McGreal organized fifty Thai students, who worked at the airport and took notes on the poor welfare of smuggled primates, and the result was a ban on primate smuggling in Thailand. The discovery in 1976 of "The Singapore Collection," an underground railroad whereby primates were transported out through Singapore for sale, resulted in the end of this racket, as well. The same year also saw a ban on primate exports in India. Malaysian exports of primates were banned after publicity broke about military training involving the animals. [505]

Throughout the years, IPPL staff members also exposed and rescued many primates living in unsuitable conditions throughout the world. The organization has fought to incriminate gorilla poachers in Cameroon and Nigeria, exporters of proboscis monkeys in Indonesia, and additional smugglers in Germany, the Philippines, Vietnam, China, and Nepal, while also securing increased protections for primates in general, including zoo primates in Cuba and Poland and chimpanzees in Saudi Arabian pet stores.[506]

If one were to color-code a global map with the locations where IPPL has made changes in the lives of primates, the resulting rainbow would truly be impressive. Still, these remarkable changes brought about overseas did not mean that IPPL neglected the welfare of primates in the United States. In 1980 IPPL oversaw the closing of a US military laboratory that conducted lethal experimentation, and it later joined forces with PETA to expose additional unethical laboratories and their unsavory practices in animal research.[507] Throughout the 1980s and 1990s, animal smuggler Michael Block was studied and his actions documented in order to successfully incriminate him in cases of illegal importation of primates into the United States. He was eventually sentenced to thirteen months in prison for his involvement in The Bangkok Six case (orangutans who were discovered in Bangkok Airport being smuggled illegally). Other smugglers were implicated, as well.[508]

IPPL has made incredible strides since its inception, proving that dedication and a strong will can truly overcome seemingly insurmountable odds. Dr. McGreal's passion and myriad successes have certainly been recognized. Among other awards, she was included on the United Nations Global 500 Honor Roll in 1992, she gained special recognition from Prince Philip of England in 2003, and she was named an Officer of the Order of the British Empire by Queen Elizabeth II in 2008.[509]

In addition to its own gibbon sanctuary in South Carolina, which has been in operation since 1977, IPPL helps support sanctuaries around the world, including the Limbe Sanctuary in Cameroon, Tacugama chimpanzee sanctuary in Sierra Leone, and others in Asia, Africa, and South America. As of this writing, various investigations and prosecutions are being pursued involving the following cases: the illegal smuggling of baby monkeys to Chicago in 1997, Egyptian authorities drowning a captured baby gorilla and chimpanzee in a vat of chemicals, Indonesian gorillas sold to Taiping Zoo in Malaysia with falsified paperwork claiming they were captive born, and the living conditions of a zoo in Thailand that keeps primates such as orangutans, chimpanzees, gorillas, and monkeys in deplorable conditions. With group members living in 31 different countries, and its advisory board members' many years of experience living and working with both wild and captive primates, IPPL is considered a great power in the field of primate protection.[510]

Born Free Foundation

The Born Free Foundation was begun in England in 1984 by actors Bill Travers and Virginia McKenna and their son, Will. However, the events that would eventually lead to the development of The Born Free Foundation actually started in 1966, when Travers and McKenna starred in the film *Born Free*, the true story of how they raised a wild lion that later had to be reintroduced into its natural environment. Their next movie, *An Elephant Called Slowly*, starred a young elephant named Pole Pole. After filming the movie, Travers and McKenna were dismayed to learn that Pole Pole had been taken from her home in Kenya and sold to the London Zoo. She lived fourteen lonely and stressful years in the zoo, without any companions or exercise, eventually dying at the young age of sixteen years old. After witnessing the untimely death of their fellow performer, the actors were determined that her life be not lived in vain, and they created Zoo Check, an organization that would eventually become the Born Free Foundation.[511]

Zoo Check began as a system of monitoring the welfare of animals living in zoos, with the ultimate goal of ending zoo institutions completely. With the creation of The Born Free Foundation, Zoo Check's work was continued to eventually encompass campaigns involving increased legislation and reporting to expose and improve upon any situation resulting in animal suffering for the purpose of human entertainment. Current programs in place include the Traveller's Animal Alert campaign, a watchdog group to which people can report animals being abused or misused in foreign countries, and the EndCap coalition, which establishes communications between the myriad animal rights groups throughout Europe.[512] The ultimate goal of the Born Free Foundation is that one day human society will end the practice of using animals in circuses, zoos, and film.

Sister foundations have also been formed. Primate-specific sister foundations include Born Free Kenya, which fights to protect Kenya's natural resources and land and to raise awareness of the illegal bush-meat trade, and Born Free USA, which merged with the Animal Protection Institute (API), a national humane organization founded in 1968.[513] Born Free USA assumed management of API's primate sanctuary. Now called the Born Free USA Primate Sanctuary, the site houses over 500 macaques, vervets, and baboons near San Antonio, Texas.[514]

Born Free's website mission statement says that they are "devoted to compassionate conservation and animal welfare. Born Free takes action worldwide to protect threatened species and stop individual animal suffering. Born Free believes wildlife belong in the wild and works to phase out zoos. We rescue animals from lives of misery in tiny cages and give them lifetime care."[515]

As evidenced by the work conducted at Born Free's various international sites, nonhuman primates have long been a focus of the organization. As of this writing, current primate-related projects include raising awareness of and putting an end to exotic pets and the use of animals in entertainment. Because new legislation is key to achieving a true difference in the lives of captive primates, the Born Free USA website includes a map with pending state and federal laws, presented clearly and effectively so that people can click for instructions on how to influence their local lawmakers.

Animal Legal Defense Fund

The Animal Legal Defense Fund (ALDF) was founded in California in 1979 by a group of attorneys who wanted to fill a void in the current legal system. At the time, there was a glaring lack of organized effort to provide assistance with legal issues involving animal rights. The founding members worked to establish a legacy that makes a true difference to those navigating the legal system on behalf of animals.

ALDF currently has offices in San Francisco, California, and in Portland, Oregon. Its members are dedicated to ending the abuse of animals by filing lawsuits against transgressors, offering free legal advice to those fighting cases to end animal cruelty, and encouraging governments to develop and uphold increasing protections for animals.[516]

ALDF works with law enforcement and animal control officers, as well as through direct involvement in the legal system. Their services are available to prosecutors working on cases involving animal welfare, and they may be of assistance during every step of an animal law case. ALDF attorneys will file litigation against those guilty of animal abuse in the hopes that, with each case bringing media attention to the plight of animals, the public will grow increasingly aware of the urgency of supporting rights for beings that are unable to defend themselves. In an introductory video on their website, Founding Director Joyce Tischler explains, "The core problem that animals face is that they are treated in our society as if they are things. Their status in the law needs to be changed from the status of 'things' to the status of sentient beings who have interests, whose interests should be protected."[517]

The group understands the importance of education and continuing to fight legal battles on behalf of animals, so it has established chapters of the Student Animal Legal Defense Fund (SALDF) at various law schools in the United States and Canada. Through SALDF, students can explore the opportunities available in the animal rights legal field through networking with professionals, attending speaking engagements, and using the literature assembled over the years by ALDF.[518]

ALDF's Animal Law Program was created in an effort to more completely integrate animal law into existing law curricula. Working with students and school officials at various universities, ALDF ensures that animal law courses are built and strengthened so that their reach may be maximized. The program also benefits those already in the legal field who are looking to organize associations of people interested in working for animal rights, and lawyers are contacted who may agree to assist *pro bono* as cases arise.[519]

The library of literature assembled by ALDF is always available for those interested and includes "model laws, pleadings and briefs, and current animal protection laws."[520] Grants are offered to fund projects that relate to animal law, and seminars and workshops help educate people about the history and future of animal law.

Although many of the group's legal victories pertain to animals in general or specific non-primate species, such as dairy cows suffering in factory farms and dogs over-bred in filthy subterranean conditions, a number of cases pertain to nonhuman primates. From 1983 to 1984 they worked to bar the importing of 71,500 monkeys into the US to be used in medical research. In 1991 they successfully sued the USDA for failure to provide adequate protection standards for the primates that were living in research facilities. In 1995 assistance from ALDF helped to stop 175 retired Air Force chimps from going back into research. One year later, they again sued the USDA, this time for failing to adequately apply the standards of the Animal Welfare Act to primates in research facilities and roadside zoos.[521]

In 1997 ALDF joined forces with The Great Ape Project, forming the Great Ape Legal Project, a "campaign to win legal rights for great apes, including the right to life, liberty, and freedom from torture."[522] As part of a project called No Reel Apes, ALDF's commitment to ending the use of primates in entertainment, a 2006 lawsuit concluded with the retirement of three chimpanzees in the possession of animal trainer and accused abuser Sid Yost.[523]

According to information published by P. Michael Conn in 2009, more than half of law schools in the United States now provide animal law courses, and more than 20 percent more US law schools address animal rights through the existence of student groups.[524] The Animal Legal Defense Fund fulfills a very real need to investigate and convict those found to be handling animals wrongly. Their educational programs, libraries, and various other means of supporting those navigating the legal system on behalf of animals ensures that an animal rights or welfare case will never be lost for lack of information.

Jane Goodall Institute

Writing about issues pertaining to nonhuman primates is nearly impossible without mentioning the work of Dr. Jane Goodall. In the fifty years of her dedication to helping chimpanzees, Goodall's influence on the way humans observe, describe, and think about primates has revolutionized primatology.

Goodall's work would never have been possible were it not for the paleontologist, archeologist, and anthropologist Dr. Louis Leakey. A white man raised within a Kenyan tribe, Leakey was instrumental in the discovery of fossilized remains proving man's origins on the African continent. Not surprisingly, his inquiries into man's history led him into primatology, and he sponsored many studies of African primates.[525]

In 1960, 23-year-old Goodall left her hometown in England to join Leakey's expedition in Tanzania. She had first met Leakey during a trip to a friend's farm in Kenya and had impressed the venerated primatologist with her passion for animals. Although she had no practical experience in primatology, Goodall had always been fascinated by the concept of living amongst wild animals, and she quickly became enthralled with the subjects of her research. The chimpanzees of the Gombe National Park would become her life's work, a dedication that would define her as a keystone defender of primate rights and ecology in general.

Over the years Goodall's persistent, objective observations have revealed many insights into the daily practices and complex lives of chimpanzees. Her habit of writing about and considering the chimpanzees as distinct individuals with assigned names, possible desires and interests, and social networks complete with varying strata of comrades and enemies created much discussion and disagreement within the scientific community. Despite her detractors' cries of anthropomorphism and excessive sentimentality, Goodall's observations were able to show human beings just how similar to their own is the behavior of African great apes, an element crucial to the fields of primatology and ecology.

For instance, when she observed two chimps sharing a meal of a baby bush pig carcass, the world learned that chimpanzees were not vegetarian (as had been previously assumed).[526] When she observed chimps fashioning wands out of plant stems and using them to fish termites from the tiny holes of their dirt homes, man was suddenly no longer the only being who could be said to use tools. This finding revolutionized the concept of human behavior. Dr. Leaky, upon hearing of this chimpanzee tool use, replied to Goodall's findings with his now famous declaration, "Now we must redefine 'tool,' redefine 'man' or accept chimpanzees as humans."[527] Other observations led to the documenting of infanticide, cannibalism, warfare, and altruism occurring naturally in chimpanzee society.[528]

The body of Goodall's research clarified many other details of chimpanzee life, including their foraging patterns, affiliation networks, dominance hierarchies, and child-rearing practices.[529] The success of her research and popularity of the cause made possible the building of a permanent research station. The Gombe Field Research Station exists to this day and is one of the longest-running wildlife research programs.[530]

Jane Goodall has built an empire of empathy for our close cousins, comprised of books, publications, film specials, conferences, and a speaking tour that takes her throughout the US and to other countries to talk about why chimpanzees deserve protection and respect.

The Jane Goodall Institute (JGI) was established in 1977 to carry on Goodall's work and, through education, to bring it to children in local communities, making a difference in the future of animals as they coexist with humans on this planet. Outside organizations working in tandem with Jane Goodall's have begun reforestation efforts so as to better protect the ecosystem and the chimpanzee groups living alongside Lake Tanganyika at Gombe, and also to link the Gombe site to other forested areas nearby that had previously been isolated due to land clearing by local people.[531]

After observing rampant abuse of primates in African communities involving the bush-meat and pet trades, a JGI sanctuary was established in Democratic Republic of Congo to ensure that there would be a healthy home available to chimpanzees confiscated from illegal activity. Projects include attempts to reintroduce rehabilitated orphaned chimps into their native environments, and supporting projects that benefit local human communities.[532] By improving the standard of living for humans who live among wild primates, JGI hopes that people will be less likely to mistreat the animals whose homes they share. Education programs teach the local people about the importance of biodiversity and ecology, and children's programs receive special focus, in the hopes that a brighter tomorrow will mean more peace for the chimpanzees of Gombe.[533]

Borneo Orangutan Survival

In the early 1990s, a Danish stewardess in her early thirties went on a vacation to Borneo. While there, she volunteered her time caring for orangutans. This experience was not only life-changing for the stewardess, Lone Dröscher-Nielsen, but would prove to be monumental for the world of native orangutans as well. Because of that trip, Dröscher-Nielsen was driven to commit her life to the native orangutans, and in 1999 she established the Nyaru Menteng Sanctuary in Kalimantan, Borneo. Now the world's largest primate conservation project, the sanctuary is home to more than 650 orangutans.[534]

As is the case with other great apes, the traits that make orangutans so fascinating to humans are also, sadly, the reasons for their endangerment as a species. Orangutans currently live in a very small area of the world (although prior to human expansion it's believed that orangutans had ranges throughout Asia).[535] The majority of orangutans now naturally inhabit only two places: Sumatra (home of the genus *Pongo abelli*) and Kalimantan (home to *Pongo pygmaeus*), which is the Indonesian part of Borneo.[536] This means that any threat to the orangutans' home ranges is also a direct threat to their survival. Any natural disaster occurring in Borneo and Sumatra decreases the likelihood of orangutans surviving as a species in the world as a whole.

Orangutans are highly intelligent, which makes them a desirable subject for humans to exploit for entertainment purposes and to keep as pets. They have accommodated the limited resources available to them in the wild and can live arboreally or terrestrially, in mountains or swamps, nomadically or statically, and have a wide and varied diet. The local culture in Malaysia considers the orangutans to be wise and worthy of reverence as old souls of the forest. There is even a myth that orangutans are actually ancient humans who hide their ability to speak so they can avoid being used as laborers.[537] These descriptions and long-held beliefs came from people who had daily interactions with orangutans and were no strangers to the ways of this great ape.

Although, overall, there are other primates more closely related genetically to human beings, orangutans have the most in common with humans when it comes to gestation period, dentition, sexual physiology and behaviors, hormone levels, hair patterns, mammary gland location, and thought patterns, such as having a longer attention span and the practice of thinking before acting.[538] There is no definitive explanation for so many significant similarities with humans co-existing with such extreme genetic differences, although it is agreed that orangutans have evolved from the common ancestor once shared between humans and other primates, and thus they are a type of preserved relic of our past. They can tie knots, recognize themselves in mirrors, and use tools to create new tools. Often more interested in the process behind solving a puzzle than the actual reward waiting at the end, they have been called "the 'mechanical geniuses' of the primate world," with "an interest and ability in solving mechanical problems...second only to humans."[539]

As is the case with other great apes, the only way for people to capture an orangutan is usually to kill the mother and take the infant. Although they do not live in the complex social groups that are seen with other great apes, orangutan infants do naturally enjoy an extended childhood and form strong bonds with their mothers, and the trauma of a capture and sudden absence of a mother means a disrupted emotional life for an infant. It's the loss of the baby's sole source of comfort, education, and chance for survival in the wild.

The effect of suffering a loss of habitat, combined with widespread killing of adults of the species, has resulted in a large number of orphaned orangutans. These parentless babies would not be capable of surviving on their own. The activities of Borneo Orangutan Survival International, the umbrella organization started in 1991 by Dutchman Willie Smits that helps to operate the Nyaru Menteng sanctuary (among its other activities), have given a future to the orphaned orangutans and recreated a safe home in their native lands.

The Nyaru Menteng sanctuary includes a rehabilitation center for orphaned orangutan infants. The naturally lush surroundings also incorporate a developing fruit plantation, and this project promotes educating the public while still allowing the orangutan infants to develop natural habits that will assist their future reintroduction to the wild. Another sanctuary run by Borneo Orangutan Survival (BOS) is located in Wanariset, where there is more focus on caring for orangutans that have been confiscated from people who possessed them illegally.[540]

Other ongoing projects are not directly tied to orangutans but still make their lives better indirectly through ecology and conservation efforts. The Mawas Program was established in order to protect the rich peat ecosystems of the area, to be used in Carbon Offset Agreements. Local communities will be taught the importance of conserving this natural area and how it will benefit their lives in the long run.[541] Samboja Lestari, the location of a former rainforest that humans had stripped, has been claimed by BOS, and its staff is working to create a rainforest that will soon house rehabilitated orangutans.[542]

The fact that the Nyaru Menteng sanctuary not only houses and protects orphans, but also attempts to reintroduce them into the wild, is unique in the field; they are not only caring for individuals who have been stolen from their ecological homes but are working to help those individuals one day return to their normal lives. The orangutan infants are fed foods that would naturally occur in their home ranges and given opportunities to develop the necessary skills and practices they would normally be expressing. This means they will not suffer from culture shock when they leave the safety of the sanctuary and start living on their own in the forests, and thus they are more apt to survive.

With an education center at Nyaru Menteng, BOS is educating the local people and school children about orangutans as a species, as well as what they need for survival in the long term. This means that future generations of local people may be more cognizant of their land-use practices and may consider that large numbers of orangutans are relying on a dwindling amount of rainforest.[543] Reaching out and educating the local community is a key step in conservation success, and rehabilitating the orangutans who have already been harmed by human activities is just one way that Borneo Orangutan Survival tries to improve the difficult road ahead for orangutans still living in the wild.

The Dian Fossey Gorilla Fund International

When Dian Fossey began her fieldwork in the late 1960s, the mountain gorillas of Rwanda were dwindling in number due to rampant poaching and human encroachment on their home ranges. Like Jane Goodall, Fossey was a student of Louis Leakey, and, like Goodall, Fossey's patient and unassuming manner allowed the gorillas living in the wild to slowly habituate to her presence. She was allowed glimpses into their life that proved to be rife with information about gorilla society that had never before been understood. While successful in bringing about change, her active fight against poachers and the society that encouraged their illegal activities resulted in her untimely death. Her murder was never solved, but many assume that it was retaliation by poachers who resented her activities, which threatened their business.[544]

What grew into what is now The Dian Fossey Gorilla Fund International started as various projects over the years, failing to be stopped even after her sudden death.[545] The Karisoke research center expanded, with cabins built to house researchers and students who would continue Fossey's work. Notable achievements of Karisoke include census data showing the various rises and falls of the mountain gorilla population over time. During the civil war in Rwanda in the 1990s, the center was looted and buildings destroyed numerous times, causing many evacuations and rebuilds and an eventual relocation.[546]

Current projects at Karisoke are numerous. Technological advances allow for imaging to track the native gorilla populations as they move throughout their home ranges. Programs encouraging ecotourism and biodiversity have begun, and golden monkeys of the region are now being habituated for study, as well. Dian Fossey saw the link between local communities and the survival of the mountain gorillas, and her fund continues to educate, support, and collaborate with the Rwandan people so as best to respect and protect the gorillas that live in their midst.[547]

The Gorilla Rehabilitation and Conservation Education center (GRACE) was formally built in 2010 in Congo, and was the first of its kind in east central Africa. It has space for up to thirty rescued Grauer's gorillas to live in natural social groups over 250 acres of natural habitat. Here injured, scared, and traumatized gorillas can receive medical attention and live peacefully and as naturally as possible as they regain their health and the skills necessary for re-release into the wild. GRACE is helpful not just for gorillas that have already been confiscated from poachers but will also help in the future, as evidence shows that government officials are more likely to confiscate illegal animals if they know there is an appropriate place that can take them.[548]

Dian Fossey Gorilla Fund International supports park officials in Democratic Republic of Congo (DRC)'s Virunga National Park in the Virunga mountains—home to Mountain gorillas—with funding, training, and educational programs. Since 2000, additional support has been given to other communities in DRC where Grauer's gorillas reside. Long-term solutions to poverty and land encroachment, such as sustainable subsistence farming, have been explored in an attempt to maintain and improve the levels of protection for the native gorilla populations.[549]

Dian Fossey's life was cut short, but her dedication and work have truly protected the gorillas she cared so much about. The fact that native Rwandan gorilla populations stopped declining and actually increased nearly 50 percent (from 260 to 380) since her research began shows what a difference she made in the gorilla world. Her research sites have provided over forty years' worth of data on gorilla life, and local communities have been enriched and supported through her legacy and the fund created in her name.[550]

The British Union for the Abolition of Vivisection / People Against Chimpanzee Experiments

A merger in 2011 brought together two groups that had long been fighting parallel battles to end vivisection on nonhuman primates.

More than a hundred years earlier, Frances Power Cobbe, a women's rights activist and philanthropist, grew disgusted after viewing animal experiments during a trip through Europe and immediately began a lifetime of animal activism, promoting the idea that "the good of mankind does not justify the 'torture' of animals."[551] Since she founded it in 1898 as the British Union, the organization that was later called the British Union for the Abolition of Vivisection (BUAV) has long since made a name for itself in the field of international animal rights. What began as the joining of five smaller societies grew into what is now a formidable force working to end vivisection completely via activism, legal battles, investigations into laboratory conditions, and promoting cruelty-free alternatives to traditionally animal-tested goods. Recent primate-related investigations involved exposing conditions of the primate trade in Cambodia and successfully encouraging the Malaysian government to prohibit export of macaques.[552]

In 2011, BUAV merged with a group called People Against Chimpanzee Experiments (PACE), a group from the United Kingdom who had been fighting to end chimpanzee research since the 1990s. Of particular interest to PACE was closing Europe's last chimpanzee research facility, situated in the Netherlands. In 2002, after years of investigations and reports of conditions at the research facility, raising public awareness and outrage, PACE and the other members of the Coalition to End Experiments on Chimpanzees in Europe tasted success when the Dutch government banned chimpanzee research and closed the facility for good.[553] Even more recently, BUAV and PACE helped lobby the European Union's ruling on Directive 86/609, which banned the use of great apes in medical research under normal circumstances.[554]

With the joining of BUAV and PACE, the groups vowed to continue their fight against vivisection and work alongside other anti-vivisection groups such as The New England Anti-Vivisection Society in the hopes that in the future, primates and other animals can live lives outside laboratory walls.

New England Anti-Vivisection Society

In 1890, Joseph Greene, a man who won an essay contest with his entry entitled "Why I Am Against Vivisection," approached lawyer Philip Peabody, one of the judges of the contest, with the intention of starting a Boston area anti-vivisection society. A group was quickly formed and became active in publicizing the truth behind medical research, with the confidence that if people learned what truly went on in laboratories, they would fight to end the practice of vivisection.[555]

Although the group fought on behalf of all animals involved in medical research, some of their projects are specific to nonhuman primates. Project R&R: Release for Chimpanzees in US Laboratories focuses on the approximately one thousand chimpanzees currently living in labs in the US (the last large-scale user of chimpanzees in the biomedical community), although the goal of the project is ultimately to end research on all great apes throughout the world. Project R&R publicizes the life stories of selected chimpanzees, so people can see the horrors of being trapped as an infant and spending one's life alone behind bars, subjected to unending painful prodding and procedures. Recent campaigns include ensuring that the remaining chimpanzees living at the former Alamogordo Primate Facility in New Mexico are retired to sanctuary life as soon as possible, and helping to place elderly primates retired from research at the Yerkes Regional Primate Research Center in sanctuary.[556]

Orangutan Foundation International

What was started in 1971 by Dr. Biruté Galdikas and then-husband Roger Brindamour in Tanjung Puting National Park, Borneo, as the local Orangutan Research and Conservation Project has grown into a dominant presence in the world of primatology and the premier destination for orphaned orangutans. Galdikas, a psychology, zoology, and anthropology student from University of California, Los Angeles, convinced anthropologist Louis Leakey (who also mentored primate protectors Jane Goodall, Dian Fossey, and others) to help fund studies of the orangutans in what was then the virtually impenetrable wilderness of Borneo.[557]

Galdikas and Brindamour began by using peaceful, culturally acceptable methods of rescuing orangutans from lives spent captive in Borneo, either living as pets or held by the government for various reasons. The organization worked in conjunction with Indonesian officials to help patrol the park grounds and fight against illegal practices such as poaching and logging.[558] As what eventually became Orangutan Foundation International (OFI) gained respect and power, Galdikas took over fully, organizing the group's efforts to become most effective in the fields of conservation, rehabilitation of orphaned or injured orangutans, research, and education.

OFI's projects have collected more than 100,000 hours of data on the activities of three generations of wild-born Bornean orangutans. At any given point, OFI's camps are home to primatology and anthropology students studying not only orangutans but other primates that live in the area, including gibbons and monkeys. A sister OFI organization has been established in the United Kingdom.[559]

Camp Leakey, named after the esteemed primatologist who mentored Galdikas, is in Tanjung Puting National Park in Borneo, managed as a joint effort by Galdikas' OFI and Indonesian officials.[560] At the site, students from around the world can observe native orangutans, proboscis monkeys, gibbons, and leaf-eating monkeys. OFI has purchased hundreds of acres of forest and has plans with locals to manage a thousand more acres and protect them from disappearing.[561]

Although various primates can be found living within the forest of Camp Leakey, its real stars are the orangutans that made Galdikas famous. In an effort to end commercial trade in orangutans, aid orangutans who are injured or orphaned due to interactions with humans, and educate local communities, Galdikas and her organization want the plight of orangutans made as visible as possible.

Currently home to more than three hundred orangutans, the center operates its own veterinary hospital and hires hundreds of people to maintain the property, provide mental enrichment to the inhabitants, and also to hold the orphaned infants 24 hours a day. The orphans, if healthy, are then released back into the forest when they are between 8-10 years old, once they have learned the key tasks for survival in the wild.[562] Feeding stations welcome the orangutans whenever they wander by, helping to compensate for the effect dwindling native land has on the natural food supply available to them. Many go on to thrive and reproduce successfully, an amazing feat when one considers how their lives in captivity might have forever altered their viability in forest living.

Although injured and orphaned orangutans have had no choice but to experience the traumas that occurred in their lives, Galdikas believes orangutans should have choices of how to live the remainder of their lives, reflected in the multiple living areas in her rehabilitation site at Camp Leakey. "We have never forced them to go back," she explained. "Why should an animal have to go back to the wild? We have to deal with the great apes as people. It should be up to them to make the decision, if they want to go back or not. To give the orangutans a choice is very important for me."[563] Thus far, more than four hundred orangutans have been rehabilitated and released from Camp Leakey, and Galdikas has been recognized for her lifetime conservation and primatology achievements with awards including Kalpataru Hero for the Earth, given for outstanding environmental leadership by the Republic of Indonesia, the distinction of Officer of the Order of Canada, a United Nations Global 500 Award, and the Pride of Lithuania Award.[564]

Sanctuaries

Primate sanctuaries are needed in developed areas and in undeveloped native lands, as well. As long as the world's human population grows and forest space dwindles, as long as science continues to expand its list of procedures and products that require testing on living subjects who then outlive their usefulness in the research lab, and as long as the viewing public laughs at chimpanzees wearing clothes, grimacing, and riding bicycles, there will be monkeys and apes who need sanctuary.

The reason(s) a primate ends up at a sanctuary may be tragic (e.g., it was orphaned in Africa after its mother was killed by poachers, or abandoned in the United States after someone changed his mind about having a simian pet) or uplifting (e.g., a Hollywood animal trainer decides to stop using primates in his act). In any event, sanctuaries are a necessary part of creating better lives for the world's primates, because they provide homes for animals that essentially have nowhere else to go.

As mentioned previously, most primates are considered "useful" by human societies solely as youngsters, when they are at the perfect but transitory stage of being smart enough to train but not yet rebellious enough to assert themselves. This means that they almost always grow stronger, less compliant, more violent, and too independent to continue on in the roles they were originally purchased or bred to serve, be it entertainer, pet, or research subject.

Some species of primates can live more than fifty years. Although at times primates can be "recycled" from one research lab to another without skewing test results, this is rare, and years of living unnatural and mentally unstimulating lives in labs can create lasting mental problems for the subjects, such as depression. Mental problems are less visible and may not be as obvious as physical health issues, but this is something that can greatly affect the way primates react in future testing situations, and it must be considered in decisions to transfer a primate from one research program or lab to another.

Luckily, for the primates who find themselves in such an unfortunate period of transition, sanctuaries now exist in almost every part of the United States and in many other countries throughout the world. Some grow out of a need to house primates indigenous to a given area, such as Borneo Orangutan Survival's Nyaru Menteng or Jane Goodall Institute's Tchimpounga in Democratic Republic of Congo. In these cases, the primates have often been orphaned by poachers or are actively hunted for the bush meat trade. Sanctuaries in these areas can help promote responsible tourism, and they also teach locals to appreciate and learn more about the species that live around them. Other sanctuaries are developed in areas where primates do not live naturally, such as Save the Chimps in Florida, and the Primate Rescue Center in Kentucky. Both facilities care for primates that have served in medical research, as well as many animals that lived as pets until they were either relinquished by their owners or taken from them by animal control officers.

Human involvement in the lives of wild primates often occurs as a sort of disaster relief: giving medical attention or caring for orphaned young. There are differing schools of thought on the ethics of human beings raising nonhuman primates. Some people would argue that this only prolongs the harm already inflicted by human beings in the primates' lives, and that they should be left alone in the wild. Others, such as primatologist Biruté Galdikas, argue that providing such care is necessary and is our obligation as humans who have inflicted harm on the creatures, "even if that means using human surrogates."[565] She believes that the bond developed between an orangutan infant and its surrogate parent is a necessary part of its healthy development, a theory that has caused conflict within the scientific community. Some research has shown that orangutans raised by humans have a worse survival rate than orangutans raised in captivity by another orangutan,[566] but unfortunately surrogate mother orangutans are not always available to care for orphaned infants, especially not in sanctuaries such as Camp Leakey and Nyaru Menteng in Borneo, each home to hundreds of orphaned orangutans. Most experts find it nearly impossible to reintroduce ex-pet primates into the wild, regardless of methodology. They are simply insufficiently prepared mentally and physically to deal with the complexities of their natural world.

After reading any of the histories of primates currently housed in sanctuaries, it may quickly become apparent to even the most hardened skeptic that these animals have suffered. Some of them never saw grass for the first twenty years of their lives. Some had never seen another member of their species before arriving at a sanctuary. Others have permanent physical marks that reflect their past: scars from lobotomies, stress-induced hair loss, or diabetes from being fed unhealthy diets as pets. Primates are very intelligent, and language studies have proven that they can remember events and even minute details from their past; these skills and abilities would indicate that the sad, sometimes vacant, sometimes wandering gaze seen in the eyes of sanctuary primates is symptomatic of their life experiences.

There's a school of thought that considers sanctuaries to be reputable only if they exist solely for the protection and well-being of their inhabitants. Proponents believe that allowing the public into a sanctuary to observe animals, even from a distance, is intrusive to the animal inhabitants, and it is therefore generally frowned upon. Others believe that any sanctuary providing a healthy home for primates is okay, even if the sanctuary takes advantage of fees charged outsiders to visit. They reason that it is increasingly difficult for nonprofit organizations to obtain financial support, and monkey chow doesn't come cheap.

It's important to remember that organizations that own primates for display or education (such as zoos, parks, and sanctuaries) are either self-governing or operate under the guidance of a shared code of behavior established by a mutually agreed-upon ruling body. A number of organizations have been established in response to common concerns observed in the fields of animal care, research, and the keeping of animals in captivity. As sanctuaries continue to be established and to exist, with taxed resources, it is clearly necessary that methods are established by which to organize and hold them to a certain level of accountability. The growth of medical research and laboratory testing means that more and more primates are being used and then displaced by the research community. Additionally, the exotic pet trade has been aided by the connectivity of the Internet, with the result that, although it is much easier for a person to find a primate to keep as a pet (legally or not), it is highly likely that at some point in the future that pet will cease to be cute and cuddly, and the maturing primate will then be turned over to live in a sanctuary. Only in the past twenty years have watchdog associations been developed to monitor and regulate the condition of primate sanctuaries in the United States, the same time period that saw an explosion in the need for and growth of American primate sanctuaries.

Since 1998, the American Sanctuary Association (ASA) has been accrediting sanctuaries that comply with their published standards of care and requirements of satisfactory organization governance. Comprised of board members who each have experience running various animal sanctuaries, the ASA is a way for sanctuary directors to work with their peers in order to reach agreed-upon higher standards of care. They encourage membership so sanctuaries can link together and share mutually beneficial information and experiences. Personal visits ensure that sanctuaries are indeed complying with regulations and are worthy of ASA approval and endorsement. ASA sanctuaries are prohibited from trade or breeding of their animals. Additionally, ASA focuses on placement of animals that are abandoned or seized by enforcement agencies and deemed unsuitable for release into the wild.[567]

The Global Federation of Animal Sanctuaries (GFAS) was formed in 2007 with the goal of establishing a universally agreed-upon standard of care and operation for animal sanctuaries within the United States and abroad. It absorbed some smaller sanctuary watchdogs, such as The Association of Sanctuaries (TAOS), which had been in existence since 1992 to regulate sanctuaries and determine which truly benefit the animals in their care. TAOS had created clearly defined regulations regarding what a sanctuary is and what it may or may not do with its inhabitants. A sanctuary, as defined by the TAOS manual, was "a facility that rescues and provides shelter and care for animals that have been abused, injured, abandoned or are otherwise in need, where the welfare of each individual animal shall be the primary consideration in all sanctuary actions."[568] Established standards indicated that once they entered into the sanctuary, no animals (or their by-products) were permitted to be sold, and animals could not be on exhibit or have contact with the public, either on- or off-site. Member sanctuaries were encouraged to develop plans and systems in case of emergencies, and requirements involving space minimums, dietary requirements, and enrichment programs were put in place.[569]

After TAOS merged with other organizations including the Captive Wildlife Animal Protection Coalition (an organization established in 2002 with the goal of ending the exotic animal trade and educating the public against keeping wild animals in captivity), GFAS gained experienced board members with experience in major animal welfare organizations such as Born Free USA, the Humane Society of the United States, and the World Society for the Protection of Animals.[570] Although at the time of the organization's inception there were other groups that had been started to monitor and act as liaison between animal sanctuaries, no main organization was universally relied upon. GFAS aimed to become that resource, as well as the first umbrella organization to also oversee animal sanctuaries abroad.

Sanctuaries are now accredited by GFAS and can benefit from the mentorship and aid provided by the organization in areas such as searching out new paths of funding and the placement of animals within the sanctuary network. Most important to GFAS is the hope that their accreditation process will illustrate the great need for reputable sanctuaries to be recognized as separate from other operations that may refer to themselves as sanctuaries but are really little more than roadside zoos or animal exploitation systems. Their educational programs explore areas such as animal cruelty and the vast side effects of displaced animals within our modern culture.[571]

The Association of Zoos and Aquariums (AZA) regulates the American facilities that exhibit animals to the public for the purposes of education, conservation, and ecological awareness. Their Accreditation Committee reviews the standards of care in each facility, assessing the living and sleeping quarters, social groupings, diets, enrichment activities, and any other aspect that has the potential to affect the physical and mental health of the animal occupants.[572] The facility itself must prove that it is managed well and upholds proper reporting and veterinary and environmental records, among other things. Once a facility is granted accreditation by the AZA, they are required to reapply every five years to prove that they are up to the constantly evolving standards of care.[573]

As of April 2013, there were 222 US facilities accredited by the AZA, although there are a much greater number of facilities operating without AZA approval. Outside the United States, the World Association of Zoos and Aquariums serves as a neutral observer to monitor and approve international animal exhibitions. [574]

In the United States, the Animal and Plant Health Inspection Service (APHIS) of the United States Department of Agriculture (USDA) regulates the care and welfare of animals, including those kept in sanctuaries.[575] USDA-approved sanctuaries are subject to inspections and public reports of inspection findings. However, approved facilities may refer to themselves as sanctuaries when in reality they may be for-profit roadside zoos. They may purchase animals from breeders with the sole intent of using them to make money, breed their animals to produce offspring, and blatantly sell animals to unsuspecting pet owners, all under the guise of being a sanctuary. This means that the uneducated person may not know the difference between a legitimate sanctuary that exists to protect endangered animals and a USDA-approved scheme to line the pockets of the owners.

The many American and international groups listed in this chapter work diligently to improve the lives of nonhuman primates living in captivity and in the wild. Without such visionary leaders and their supporters and staff, one can hardly imagine the dire situation that primates would face in today's world. The life's work of these primate protectors has helped thousands of nonhuman primates and at times ensured the very survival of species, proving that sometimes one person can indeed change the world.

8. Ethics of Primate Use

The notion of 'us' as opposed to 'the other,' which, like a more and more abstract silhouette, assumed in the course of centuries the contours of the boundaries of the tribe, of the nation, of the race, of the human species, and which for a time the species barrier had congealed and stiffened, has again become something alive, ready for further change.[576]

-Peter Singer and Paola Cavallieri

We are joined to the other primates evolutionarily, and this genetic proximity has tied us together for much of man's recorded history. Nonhuman primates have fascinated people since they were first viewed by humans, removed from their forest homes, and thrust into a world of observation and fascination, but also one of mockery, fear, and derision. By examining the history of our tumultuous relationship with the other primates, we may use the powers of hindsight to focus our aim on an ethically considerate future.

Currently, the oldest primate fossil records belong to prosimians and date as far back as the Eocene epoch, 55 to 34 million years ago.[577] These first primates were tree-dwelling, fruit eating mammals who would later branch off and evolve into the more adaptable anthropoids, who first appeared 38 million years ago,[578] and great apes, who were around 23 million years ago during the Miocene epoch.[579] Although the first distinctive labels that identify the primate kingdom didn't arise until Linnaeus's classification system in 1758, artifacts prove that humans had

213

been living among other primates and using them for various purposes since ancient times.

Egyptians coexisted with baboons, offering them the best food and drink in deference to the baboon god, Thoth.[580] Grecians traded vervet monkeys as pets, and African monkeys started appearing in European artwork as early as the late fourteenth century. In 1607, Andrew Battell, an English sailor who had been held captive in Africa, returned to his homeland with stories of monsters called pongos (the Latin word we now use to name orangutans *pongo pygmaeus*) that were half human and half beast: "This Pongo is in all proportion like a man, but that he is more like a Giant in stature, then a man: for he is very tall, and hath a mans face, hollow eyed, with long hair upon his browes. His bodie is full of haire, but not very thicke, and it is of dunnish colour...They sleepe in the trees, and build shelters for the raine. They feed upon Fruit that they find in the Woods, and upon Nuts, for they eate no kind of flesh. They cannot speake, and have no understanding more than a beast."[581] We know now that he was referring to chimpanzees, gorillas, and bonobos (the primate species indigenous to Africa). Afterwards, the word *ape* was often used to refer to any human-like primate observed, most often the Barbary macaque.[582]

Three years after Andrew Battell had waxed philosophical about Pongo, Shakespeare's last completed play, *The Tempest,* was written, exploring how people might treat a being that was part man, part animal. The socially adept character, named Caliban, is treated as a slave with dangerous strength who cannot be trusted...sounding a lot like a great ape of some sort. Historical records show that a live chimpanzee was received in Europe in 1640,[583] and the first accurate anatomical depiction of an ape occurred in 1641, when Dutch anatomist Nicolaas Tulp dissected the body of what he at the time decided was an Indian satyr called an *orang-outang* by locals in Africa. It's now obvious to us that it was not an orangutan, which never lived in Africa, and at first it appeared that he was describing a chimpanzee. However, further analyses by primatologist Vernon Reynolds determined that Tulp's ape was most likely a bonobo, a chimpanzee-like great ape recognizable by its toe webbing, which occurs much more often among bonobos than among chimpanzees. Additionally, the area where the specimen was procured, near the Zaire River, is a sort of dividing line between the natural ranges of chimpanzees and bonobos,[584] and the combination of physical markings and home range location allows us to believe that Tulp was actually dissecting a bonobo.

Upon dissection of the first living chimpanzee in England, who died a year after he was brought into the country in 1698, physician Edward Tyson noted that the specimen was "a sort of *Animal* so much resembling *Man*, that both the Ancients and the Moderns have reputed it to be a Puny Race of Mankind."[585] Sixty years later, in 1758, the taxonomical groupings of Carl Linnaeus were developed and were subsequently supported in 1859 by Charles Darwin's evolutionary theory in *On The Origin of Species*.[586] In 1876, David Ferrier of King's College Hospital in London published the first comparison of human and monkey brain functions, inspiring many others to complete further studies comparing primate gray matter,[587] including nonhuman primate reactions to diseases that had plagued humankind. Thus the bioresearch system as we know it today was conceived.

At various points in history, nonhuman primate species have been viewed as humorous or as symbols representing the devil, psychosis, ancestral prehistory, intelligence, violence, and sneakiness. Even some of the terminology assigned to primates has sinister etymology; for example, marmoset meant a small marble gargoyle, or any other grotesque figure. Although the word monkey didn't formally appear in the English language until the early sixteenth century, popular satire stories of the twelfth century involved ape characters named *Moncke* or *Moneke* in German, or *Monnequin* in French, all of which derive from the Spanish word for ape (*mona*) or the English word *mannekin*, a diminutive of *man*.[588]

Accounts and descriptions of chimpanzees and gorillas were created in the early to mid seventeenth century by Europeans who had seen the fantastic creatures during overseas travel. The medieval Japanese considered the native macaque to serve as a holy liaison between humans and their god, yet in more modern times, that same macaque was viewed with scorn and derision, as nothing more than a jester. Chinese culture considered gibbons to be liaisons, as well, but between lowly man and the higher ideals of poet and philosopher.[589]

In India, the monkey god Hanuman is worshipped as a symbol of service, courage, and unquestioning sacrifice of life.[590] Indigenous tribes of Borneo believed in myths that the mysterious orangutans that inhabited the forests of their land had actually once been human but had lost their voices, and folklore told that an orangutan had once kidnapped a female human and kept her as his wife. The very name orangutan means "person of the forest" in the native Malay language.[591]

The varied and complex personalities assigned to primate species throughout history make it seem as if ancient cultures recognized the ambiguity of nonhuman primates, and their delicate positioning between the human and animal kingdoms afforded them special attention. In the long-term relationship between humans and nonhuman primates, this would prove to be both a blessing and a curse.

Although the behaviors of monkeys and apes were first viewed with interest and even fascination, some of their innate tendencies soon struck fear and loathing into the hearts of an increasingly Christianized audience. According to primatologists Frans De Waal and Frans Lanting, "Early Christianity taught that sexual behavior, except in the cause of reproduction and within the bounds of marriage, was a sin... and it showed itself in minor ways too, including man's attitude towards the monkey. For those characteristics of the monkey that amused the ancients—its imitativeness, greed, and sexuality—failed to amuse the Christians; these qualities now took on sinister proportions as sins, for which the soul of man would be eternally condemned to hellfire. From being an object of fun and games, the monkey became nothing less than a symbol of the Devil himself."[592] Myths of satyrs, monkey-like half-men who inhabited remote islands, told of warfare, raping women, and a type of animalistic hedonism guaranteed to strike terror into the heart of any god-fearing mortal. It's fortunate that, throughout history, various ideals of the times have also been observed in primates, so that instead of being viewed forever as wanton lechers, they have also been praised and noted for other characteristics that make up their complex lifestyles.

Adventurers of the nineteenth and twentieth centuries aimed to explore undeveloped lands and record their cultures, flora, and fauna. Regrettably, they often plundered the natural world as they went, and as there were few ways to document their findings, the bodies and hides of native animals were often the prizes of such excursions. The explorers leading these expeditions became well known for their writings about the strange lands and their thoughts on what they found, writings that are now seen to be rife with racism, inaccuracies, and paternalistic notions of the superiority of the white man. Often, the infant animals of slain mothers were kept as trophies or pets or were shipped back and sold in America or Europe. Most of these living examples didn't survive long, due to the inexperience of their caretakers, malnutrition, injuries sustained during capture or transport, and illness.

The sense of species-centric detachment and moral superiority of the explorers is no more evident than in 1906, when the celebrated adventurist, hunter, and naturalist William Temple Hornaday not only captured an orangutan named Dohung for display in the Bronx Zoo but caged him with a human pygmy from the Congo named Ota Benga. After two years of living in captivity and being treated as a zoo display, Ota Benga was finally released.[593] Unsurprisingly the orangutan, Dohung, never lived free again.

In the way that the unknown tends to strike fear into the hearts of humans, nonhuman primates have also been considered highly dangerous or murderous. One need only refer to the movie *King Kong* to see how easily and often the peaceful gorilla, for example, was distorted into a super-sized threat. The simplistic animalism seen in nonhuman primates was also used to describe the behaviors of various people considered uncivilized by the bourgeoisie in the 1800s, such as criminals, prostitutes, and other unsavory types.[594] A Belgian propaganda film of the 1950's calmly showed the shooting death and skinning of a mother gorilla, all while her baby cried in the background, so that the baby could be added to the ranks of a zoo. There was no moral outcry over the film.[595] Luckily, by the 1960s, the results of Jane Goodall's field study of chimpanzee society would open people's eyes to the rich emotional lives of wild primates, and filming the death of a mother gorilla would then have been deemed unacceptable.

One of the subtler ways of reflecting humanity's respect for animal species is not only how we treat them but the words and phrases used to describe them. Dr. Jane Goodall learned this lesson the hard way when she started publishing her observations from the field in Gombe in the 1960s, having little experience with the rules or expected practices in scientific reports. To Goodall it was natural to refer to the busy chimpanzees as "him" and "her," not as "it," and she named the chimpanzees to keep track of the relationships, both familial and affiliate.

Goodall also tried to describe the emotions she viewed in her subjects, thinking that this would more accurately paint a picture of the chimp society she was studying. The scientific community was aghast at the methods she used to write about the chimpanzees. As she described it, "When, in the early 1960s, I brazenly used such words as 'childhood,' 'adolescence,' 'motivation,' 'excitement,' and 'mood' I was much criticized. Even worse was my crime of suggesting that chimpanzees had 'personalities.' I was ascribing human characteristics to nonhuman animals and was thus guilty of that worst of ethological sins—anthropomorphism."[596]

As time went on, other primatologists such as Dian Fossey, Shirley Strum, and Barbara Smuts would come to value anthropomorphism (or at least a careful sort of anthropomorphism, something that an article by Amanda Rees refers to as "speculative anthropomorphism"[597]) as a tool to augment their comprehension of their research subjects. In their view, much can be lost in the process of avoiding any semblance of anthropomorphism, whereas ascribing to the animal what the observer believes the animal is thinking brings a depth and better comprehension of its actions.

As Karen Strier put it, "Data alone do not convey what the day-to-day experience of accompanying muriquis [also known as the wooly spider monkey] has been like...it is the stories about the monkeys and the progress of the research that provide an essential context for the scientific findings."[598] Animal cognitive expert Marc Bekoff elaborated on the topic: "I...use anthropomorphic words and phrases that seem most likely to capture the essence of what [an animal] is doing. I could describe, in great and inconvenient detail, [the animal]'s behaviour from anatomical and physiological perspectives, but this approach would convey little or no useful information to another person about what [the animal] was doing."[599]

Also of note is the consideration that getting a glimpse of the possible emotional lives of animals, whatever the method, draws a reader in and more greatly interests readers in the animals' lives and safety, which can be crucial for endangered beings. However, there is a wide gap between blindly assumptive anthropomorphism and, on the other side of the spectrum, completely discounting the emotions of animals. In the middle is the consideration that animals have their own points of view; they experience the world, perhaps not the same way human beings do, but in a way that makes sense to them and their particular sensory receptors, and this is certainly deserving of recognition, respect, and consideration.

The speculative anthropomorphism described by primatologists and animal behaviorists is hard to avoid. The concept of evolutionary continuity would imply that, considering the recognized biological ties between human and nonhuman primates, behavioral ties would exist as well. In that light, it's not far-fetched to assume that animals may have similar thoughts, emotions, and awareness as we. Primates have been observed using many of the same body language signals as humans, such as an infant whimpering and holding out a hand in order to get attention from the mother, the pursing of lips to express disappointment, or the crossing of one's arms over the torso in an act of stubborn defiance. Great apes of many species laugh when tickled, and as mentioned previously, young bonobos have even been observed making silly faces just for their own amusement.[600]

Of course, some behaviors appear to be exactly the opposite of what occurs naturally in humans, such as the fear grin of chimpanzees and bonobos, so often mistaken for a happy smile by humans. The bearing of teeth is an obvious intimidation display in some primate species, such as baboons, but chimpanzees and bonobos happen to grimace in such a way as to appear to be mocking the human smile. Anthropologist Frans de Waal speculates that this is the origin of the nervous smile of humans, because humans and other primates feel the need to appease others when placed in a frightful situation, and so show their teeth in a smile or grimace.[601] The burden of proof may fall on those people who don't believe that primates can have feelings, beliefs, and desires similar to those of humans, for then they must provide alternate reasons to explain the actions of primates when they show behaviors humans seem to relate to easily, such as begging, mourning, crying, or playing.

Fortunately the scientific community has changed since Dr. Goodall first published her studies and, depending on the context, discussing the motivations behind animal behaviors is often the rule instead of the exception. Many primates exhibit actions that are uncannily similar to those of human beings. For instance, "Two friends may greet with an embrace and a fearful individual may be calmed with a touch, whether they be chimpanzees or humans."[602] During one particularly poignant moment, Dian Fossey saw a young gorilla cry after her family was slaughtered in front of her. As Fossey saw the tears of this creature, staring out the window at the mountains that were once home to her family, the motivation behind such behavior could hardly be questioned.[603]

And yet, two beings performing similar actions is not a fool-proof guarantee of similar consciousness when making interspecies comparisons. After all, says neuroscientist and evolutionary anthropologist Terrence Deacon, "[a]nimals can have conscious minds without sharing all the attributes of human consciousness."[604] This is probably one of the principal reasons that so many discussions of animal consciousness end in stalemate. There are no guarantees, and it's only when static determinations emerge that presumptions and hypotheses morph into facts.

There are different levels of consciousness. A higher level of consciousness, which it is generally assumed that healthy humans attain, may involve having a unique personality, likes and dislikes, and goals and phobias. If a person capable of this higher level of consciousness were suddenly disabled with a brain injury, for example, he or she might no longer be able to discuss or even be aware of these unique personality traits, but it would likely be considered morally reprehensible to inflict pain or suffering on him as part of an experiment. In such an instance, the level of sentiency would not dictate the right to live free of pain. Why then, does this rule only apply universally toward human beings? We know that most, if not all, nonhuman primates are not only sentient but also able to prove that they have a higher level of consciousness, but perhaps because they are not able to beg in a human language for it, they are not granted freedom from pain. As Marc Bekoff asserts simply, "Their own pain and suffering are no less important than our pain and suffering."[605]

It's important to remember that animals can process sensory information differently from humans, so comparing animal to human reactions isn't always practical or accurate. But as PBS's *Nova* program *Ape Genius* described, "…as the most social of apes, we can't help reading thoughts and feelings into the mind behind any familiar face."[606] Michael Tomasello, co-director of the Max Planck Institute for Evolutionary Anthropology in Germany, admits that one of the most striking strengths of human beings is their ability to identify with others on a deeper level than any other primate observed thus far.[607]

Humans are innately drawn to consider the emotions and motivations of others, even across a species barrier. This thorn in the side of anthropomorphism is hard to refute and should perhaps serve as a reminder that nothing can be assumed as fact when one species tries to look deeply into another. Unless the individuals being observed are capable of using advanced human language (which has not proved possible for nonhumans), there is no way to know for sure what they are thinking. We can only guess. With the knowledge of our extreme genetic similarities to other primates, we can only rely on inferences and our own past experiences, while also considering what we believe to be the best interests of the animal, to determine what they may be feeling.

Both field research such as Jane Goodall's and more traditional lab research through the present day have, perhaps not surprisingly, shown that nonhuman primates can be both violent and peaceful, cruel and kind, and, like human beings, extraordinarily diverse in many other respects. But while it's clear that our interpretations of nonhuman-primate abilities and intellects have varied greatly through the centuries, one thing that has not varied is humanity's generally assumed position of moral superiority over the rest of the animal kingdom. Situations where nonhuman species are held in great respect due to religious or cultural beliefs are fewer and farther between, and most often, nonhuman species are appreciated only in terms of their usefulness to humans in attaining particular goals.

It seems likely that if you were to ask a random sampling of people why they believe human lives are inherently worth more than animal lives, you could get as many different answers as there are people in your sample. As discussed earlier, many branches of the scientific community have been dedicated to finding ways to uphold humans on their pedestal of moral authority and worth. An easy path to a completely human-centered view of the universe is followed simply by failing to consider the biological relationship between humans and other nonhuman primates. For example, the field of evolutionary psychology studies ways that the human mind functions uniquely. Ignoring the links to primatology (even those that are purely physical, such as those pertaining to brain structure) leads to a disassociation of humans from other primates and, thus, indirectly reinforces the pervasive distinction and moral hierarchy.

The field of ethics addresses granting respect to living beings that innately deserve such consideration and basing one's actions and treatment of others accordingly. Sometimes, when considering proper treatment of others, agents are required ethically to consider the interests of others affected by their actions, which could mean considering the interests of other people, populations of people, inanimate-yet-living objects such as the natural environment, and the animals that populate that environment.

When a living creature understands the existence of other beings' mental states (such as desires, fears, and knowledge), it is said to have exhibited theory of mind. However, the theory of mind of a being with whom we lack a common language is easily commented upon and critiqued but difficult to prove in the concrete, absolute way that we humans prefer in our science. Primatologist Craig Stanford proposes that if an animal is observed to be acting in such a way as to suggest that it understands the mental state of another individual, it is showing evidence of a theory of mind. Clearly some of this notion is based on assumption, and whenever a human infers anything about the behavior and intentions of a nonhuman, anthropomorphism looms in the background like an unavoidable shadow.

Max Sheler, a German biologist of the 1920s, analyzed the behavioral research of chimpanzees and was of the opinion that their actions were fully dictated by their instincts and biological needs. Because they were found to be "open to the world," as he expressed it, they could not comprehend indirect references to things and abstract concepts, nor were they self-aware. The chimpanzees were only aware of things related to their direct survival needs.[608] Modern research, as has been detailed here, has proven that Sheler underestimated the depth of chimpanzees' consciousness, but even if that were not the case, a shallow consciousness does not imply that basic rights should not be granted to an intelligent individual or any living creature, for that matter.

Vivisectors have argued that primates and other animals lack a sense of mind and awareness of the future; thus, they do not suffer when spending years living in a laboratory. Assuming that primates do lack a sense of awareness as described, it's possible that this may cause them to suffer even more. As ethicist Andrew Linzey stated, "Unlike humans, who can understand the reasons for their captivity, this creature cannot rationalize his predicament. Rather, he experiences the terror of imprisonment without any softening of the experience that comes from intellectual comprehension. And, since an animal's very life depends upon the acuity of its senses, the denial of liberty to a free-ranging creature constitutes a severe deprivation."[609] This deprivation of liberty does not occur only when primates live in laboratories but can happen any time they are removed from their natural surroundings and forced into doing anything not of their own volition.

By the last decade of the twentieth century, the numerous ways that human societies relied on nonhuman primates as pets, entertainers, research subjects, and the focus of education culminated in a philosophical response by some of the greatest minds in ethics and animal behaviorism. *The Great Ape Project* was presented as both a book and a revolutionary concept. The book was edited in 1994 by ethicists Peter Singer and Paola Cavalieri and included essays written by prominent primatologists, scientists, and ethicists discussing the myriad strengths and abilities of great apes as reasons to afford them more ethical consideration in modern human cultures.

The driving force of The Great Ape Project as a movement was simple: Humans have traditionally been entitled to a "precious moral status,"[610] one in which people within the group are granted certain rights and those outside that sphere of influence are denied those rights. In relation, one of Peter Singer's earlier books, *Animal Liberation*, coined the term speciesism, which is an unfair assignment of rights based purely on species. As he described it, speciesism "is a prejudice or attitude of bias in favor of the interests of members of one's own species and against those of members of other species."[611]

Animal Liberation promoted a more equal consideration of the interests of other species. For example, although a person may believe that animals deserve lives free of intentionally inflicted pain and suffering, that person would most likely not have the same concerns about rocks, because rocks lack a nervous system and presumably cannot feel pain. Yet, despite the beliefs of some people that animals don't feel pain, empirical evidence proves that they do. Animals not only have nervous systems, they have the same physiological reactions as humans when exposed to what humans consider to be painful stimuli, such as increased blood pressure, dilated pupils, and perspiration, and they also exhibit similar behaviors, such as avoidance of or attempts to avoid pain-producing stimuli. The diencephalon, the part of the brain where impulses, emotions, and feelings are produced, is also well developed in species other than humans.[612] Such parts of various animals' anatomy have evolved just as in humans and have contributed to the survival of those species in the Darwinian sense; thus, it is illogical to suppose that pain felt by animals is any less severe or uncomfortable than pain felt by humans. Furthermore, many animals have more acute sensory receptors than do humans, which may mean that they may suffer even more pain than humans do.

Similar to the ways sexism and racism were questioned and are now thought of by many as arbitrary, prejudiced, and wholly without intellectual merit, *Animal Liberation* raised questions about why humans might be entitled to greater protections than other species. As Singer explained, "Racists violate the principle of equality by giving greater weight to the interests of members of their own race when there is a clash between their interests and the interests of those of another race. Sexists violate the principle of equality by favoring the interests of their own sex. Similarly, speciesists allow the interests of their own species to override the greater interests of members of other species. The pattern is identical in each case."[613]

In the forward to Gary Francione's *Animals, Property, and the Law,* William M. Kunstler, cofounder of New York's Center for Constitutional Rights, goes even farther than merely comparing speciesism to racism; he believes that permitting speciesism to exist allows a society to more easily discriminate against disadvantaged groups of people."[614] Animal rights lawyer Steven M. Wise summed up this idea neatly with the statement, "If two classes pose the same harm or deserve the same benefit, then discrimination in favor of one class against the other is probably irrational and arbitrary."[615] These ideas and others beg the question, why should human desires be weighed more heavily than those of other species?

In most modern cultures, human desires are considered paramount, even when they conflict with what we presume to be the interests of other species. Such speciesism is seemingly easy to justify, but when the implications are examined, it becomes a slippery slope. Consider the theory not across species but across populations within the human species. Basing charity on ethnic and geographic similarity, for instance, might mean that we should not only help those closest to us on the great family tree but those closest to us on the map, as well. If we should help those that live closest to us, this negates any thought of applying our money to where it can be best used. Does the family living in poverty down the street need your money as much as the orphan living on the streets of India? No? What about an orphaned chimpanzee currently en route to a breeder in Florida, who has recently seen his mother killed, is weak and dehydrated, and has been condemned to a life of servitude and a diet that makes him sick?

One of the defining elements of human thinking and language involves the ability to compartmentalize. By being able to sort things into definable categories, we can make statements about our world and opinions. We can think abstractly despite the overwhelming barrage of colors, shapes, ideas, concepts, feelings, principles, and noises that assault our sensory organs every day. As our cultures continually evolve, we revise the compartments that we use to define our world. Sometimes we decide, perhaps as a culture, to frown upon assigning hierarchies to specific distinctions (most recently in the West, along racial and sexual lines). It's the opinion of Singer and Cavalieri that "the notion of 'us' as opposed to 'the other,' which, like a more and more abstract silhouette, assumed in the course of centuries the contours of the boundaries of the tribe, of the nation, of the race, of the human species, and which for a time the species barrier had congealed and stiffened, has again become something alive, ready for further change."[616]

It is compelling to ponder the fact that humans rationalize and justify our treatment of other species when ours is but one of the myriad species inhabiting this planet. In the context of American culture, this attitude has raised many highly debatable, even uncomfortable issues that require a rethinking of the subject of the great apes. As research has increasingly proved mental, physical, and social similarities between human and nonhuman primates, and specifically the great apes, questions are increasingly raised as to why humans have rights that are routinely denied their genetically close cousins. As Jared Diamond points out, "The genetic distance (1.6 percent) separating us from pygmy [bonobo] or common chimps is barely double that separating pygmy chimps [bonobos] from common chimps (0.7 per cent). It is less than that between two species of gibbons (2.2 per cent)...the remaining 98.4 per cent of our genes are just normal chimp genes."[617]

Singer and Cavalieri were very clear on the terms of the Great Ape Project: "We demand the extension of the community of equals to include all great apes...The 'community of equals' is the moral community within which we accept certain basic moral principles or rights as governing our relations with each other and enforceable at law. Among these principles or rights are the following:
1. The Right to Life
2. The Protection of Individual Liberty
3. The Prohibition of Torture"[618]

The authors then go on to define those principles more specifically, including that the right to life may be revoked in self-defense, detention of individuals may only occur when it is in their own best interest or protection, and deliberate infliction of pain, even if supposedly for the benefit of others, is simply wrong. If the legal toes of primates are being stepped on, a human representative should be assigned to fight for their protection, similar to the way a court may appoint representation to human beings who are unable to defend their own best interests.

As Peter Singer puts it, "Moreover, to grant rights to the great apes would not threaten any major industry nor the diets of the majority of the population, as equal consideration for all animals would…If the Great Ape Project is successful in leading us to include, for the first time, members of a nonhuman species with the sphere of beings who we recognize as having basic rights, then it will have served to bridge the gap between humans and other species. It will then make it more feasible to extend equal consideration to other nonhuman animals as well."[619]

Granted, Singer, and Cavalieri pose persuasive arguments, so why limit their book and discussions only to great apes? Why not fight for the rights of all nonhuman primates? It may be that the authors wanted to take advantage of the close biological ties between great apes and humans, perhaps because it is often hard to deny the physical and mental similarities between the two groups of primates. They may have reasoned that if they could build a case that greater legal protections should be granted to great apes, it might be possible to later include primates farther away on our shared family tree. Using such cases as precedents, it is conceivable that legal rights might eventually be granted to all nonhuman primates.

The Great Ape Project and similar ethical considerations are crucial to improving the future for nonhuman primates. Simply because their ancestors were mistreated and abused by our forefathers is no reason to continue such inhumane interspecies behavior in the future. The ways that humans use and interact with nonhuman primates in the modern world are numerous and varied, and each environment that includes nonhuman primates deserves close ethical examination.

Ask most people to express their thoughts about the behavior of monkeys and apes, and they will likely use caricatures: chimpanzees at the zoo spitting or throwing feces or cartoon-like gorillas pounding their chests, as in a Tarzan film. In such cases, it's important to remember that the behavior of animals in captivity is rarely indicative of the way they live naturally in the wild. Those manmade environments may look convincing to people, but a wild-born primate isn't fooled by faux rocks and rope vines. When animals are placed in such unnatural settings, all behavior is altered, and behavior indicative of stress or unhappiness may be exacerbated. Aberrant behavior can include overeating, rocking, apathy, catatonia, violence towards others, and even self-mutilation such as plucking of fur and biting of digits or limbs, sometimes to the point of amputation.

Current United States Department of Agriculture regulations stipulate that chimpanzees, for example, have a minimum-size cage of 5' x 5' x 7',[620] and it's often the business practice of laboratories to provide only the minimum requirements in order to make the most use of their available space. Although it's understandable that laboratories wish to maximize their profits, it's important to realize that a full-grown chimpanzee can weigh 120 pounds. Locking a human being of this size in a 5' x 5' space for the majority of his life would be tantamount to torture. When primates have been released from years of languishing in indoor labs or similar captive environments, they are often pale, sickly, and weak due to muscle atrophy from lack of space to exercise and exhibit naturally occurring behaviors.

The range of sizes and differences in behavior among nonhuman primates means that there are no universally acceptable living conditions for primates as a whole. Crab-eating macaques like to run, gibbons like to swing arm-over-arm for great distances, capuchins are climbers, and orangutans traverse very large areas in their natural home ranges.[621] There is no enclosure that could satisfy all these natural behaviors universally, and it requires much research and expertise in the intricacies of these species' lives to provide each with appropriate enclosures and enrichment activities sufficient to maintain their physical and mental health.

When the USDA determines that labs are not following the minimum requirements for primates in their care, the punishments doled out rarely include closure of the operation as a whole. This allows the mistreatment to continue and gives the research facilities the opportunity either to fix systems completely or merely to adjust living conditions so as to meet the USDA's minimum standards.

When labs close voluntarily or are shut down due to infractions of the Animal Welfare Act or other standards, the animals are considered surplus equipment that requires placement, and primate sanctuaries often find themselves suddenly facing the arrival of new inhabitants. If there is room for them, special care and funding must be secured for their transport, socialization into the existing groups, and future care.

When New York University's Laboratory for Experimental Medicine and Surgery in Primates (LEMSIP) closed in 1996, after years of undercover and federal investigations found countless instances of cruelty and failure to adhere to the legal standards of care for their primates, hundreds of chimpanzees and monkeys were suddenly homeless. Some of the chimpanzees at LEMSIP had been recycled from sign language studies, so despite having spent the first years of their lives being coddled, like human babies, they had later spent decades at LEMSIP never seeing the outdoors or touching the floor again (for sanitation purposes, they were housed singly in cages suspended from the ceiling).[622] LEMSIP primates were eventually dispersed throughout the United States network to sanctuaries such as the Primate Rescue Center in Kentucky and Chimpanzee Sanctuary Northwest in Washington.

When the US Air Force ended their chimpanzee research and donated their surplus chimpanzees to The Coulston Foundation, a New Mexico facility with a very long track record of failing to care properly for the primates in its care, the sanctuary Save The Chimps fought for ownership of the primates, knowing that their living conditions would be very poor if they remained in The Coulston Foundation's dismal buildings. After a year-long struggle, 21 chimpanzees were awarded to the sanctuary's care, a number that would grow by the hundreds when The Coulston Foundation entered bankruptcy and closed in 2002. Save The Chimps took over the Coulston Foundation's buildings in New Mexico, and although improvements had been made for the chimpanzees living there, in 2011 the nonprofit organization was finally able to transfer them all to its vast primate sanctuary in Florida, replete with outdoor living spaces on manmade islands.[623]

Even without considering the long-term care of primates who have spent their lives in laboratories, there is an underlying conundrum inherent in any defense of animal testing. Those who believe that animal testing is ethically allowable because animals are substantially different from humans must be questioned about the usefulness of testing on nonhuman subjects. Surely the results cannot be applied to humans if we are, by this definition, so very different from animals. Since great apes have been shown to exhibit mental capabilities equal to that of a four-year-old human being, the question that begs to be asked is that if it's acceptable to submit such aware simians to medical testing, why then is it not acceptable to submit a human infant to medical testing?

Nazi experiments on what they considered to be inferior people (of different races, religions, or disability levels) are now universally condemned as cruel and despicable. Will there be a day in our future when nonhuman primates are no longer considered a lesser species and are granted freedom from being our research subjects? As Deborah Blum so eloquently put it, "Chimpanzees in AIDS research perhaps serve best as a haunting example of the biological tradeoffs of being almost human."[624]

Peter Singer has developed a hypothetical, and upsetting, suggestion to use orphaned and mentally disabled human infants as subjects on which to perform medical testing. He proposes using orphans so as not to cause pain to any parents, and specifically mentally disabled orphans because they hypothetically lack awareness and a sense of fear. He postulates that this would cause the least amount of suffering (much less than testing carried out on nonhuman primates) and would have the added bonus of increased usefulness of data, since the tests would actually be carried out on humans, and interspecies differences would no longer have to be considered.

Many studies involving animal testing are questionable and thought to be superfluous and/or medically unnecessary. For instance, some that aim to find the maximum level of a substance that may be ingested prior to causing certain death are entirely cruel and seem to have little relevance in the real world. Although conducted upon other species, these tests are supposedly required to determine the level at which death or overdose would be experienced in humans.

Roger Fouts, who created waves in the scientific community when he decided to stop the decades-long language studies with Washoe and her chimpanzee companions after developing a deeper respect for his subjects, has since spoken out against animal research. In less of an about-face than an evolution of thought, he explained it this way: "If you look at the research community, somebody has got to stand up and say the king has no clothes. Chimpanzees aren't numbers; they aren't hairy test tubes. They are folks who are going to have major problems if we don't replace human arrogance with compassion."[625]

In the case when testing modules are developed to cure a true medical affliction that causes pain and suffering, it's easy to see why some people may support them, even if animal lives are involved, as the tests are likely to benefit humans in the end. Jan Moor-Jankowksi, onetime director of the aforementioned LEMSIP laboratory, explained, "Imagine if Landsteiner [one of the first to study the effects of polio by injecting the virus into the spines of two monkeys] had access to monkeys in 1908. We might have had the polio vaccine 20 years earlier. Tens of thousands of people would not have been paralyzed."[626] As Deborah Blum stated in the preface to her book *The Monkey Wars,* "Of course, we would rather have the drugs and surgical procedures and the wonderful medical advances without being told that so many hundred animals died for that information. And if we do know that animals are used, at least we would rather be spared seeing them bleeding in a medical test."[627]

Researchers have spoken out in defense of the quality of care their animal research subjects receive and are constantly having to prove the validity of their research and fight against groups such as PETA, Animal Liberation Front, and others who so eagerly would like to end their life's work. Lobbying groups such as the National Association for Biomedical Research work to improve the public image of animal testing by advocating humane and compassionate treatment of animal laboratory subjects. Using the same propaganda techniques as groups opposed to medical testing, they distribute media aimed at convincing the public that animal testing is a vital part of the research community, something that is a necessary element of humanity's continuing medical progress.

Of course, not all captive environments are as stark and obviously uncomfortable for their inhabitants as bioresearch facilities. Primatologist Craig Stanford spoke for many when he expressed this hierarchy by writing, "Keeping great apes in zoos is morally questionable, and in laboratories reprehensible."[628] Zoo proponents will explain that zoos fill a very real educational need, because they allow people to view exotic animals that they would never be able to see without traveling great distances. This in turn can foster a greater appreciation for the animals and, by extension, an increased consideration of ecology. That said, since the behaviors of captive animals are very much affected by their altered environments, living in circumstances unnatural to their species means that the public is not seeing what a chimpanzee or elephant is actually like, as these captives cannot replicate the behaviors of their brethren in the wild. In this sense, the sacrifice of the normal life of the animal on display may be in vain. Animal rights activists will often argue that keeping animals in cages only benefits the zoo itself and entertains the humans who pay to gawk at them.

In this instance, the concept of a theory of mind also comes to the forefront, as it is difficult to prove that a primate is aware of his lack of freedoms in a zoo. Some insight into this issue came from Gary Shapiro, a researcher studying language comprehension of orangutans. He spent time in the 1970s with Aazk, a young orangutan living at the Fresno Zoo in California. He and Aazk both seemed to enjoy the time they spent together in her cage, but Gary quickly noticed a change in Aazk's behavior when he left her. He described her attitude as "disappointed" and "resentful." "I didn't mind being in the cage," he elaborates. "In fact, it gave me a better perspective what it might be like to be locked up like a zoo animal. Yet, perhaps she understood that my time in the cage was temporary while she had to stay in the cage."[629]

The quality of zoo enclosures varies widely from country to country, depending on local attitudes and funding available for facilities and upkeep. One of the main concerns of animal rights groups is substandard care of animals in both accredited and non-accredited, or "roadside," zoos. The proliferation of such businesses and relative difficulty of accurately monitoring their facilities means that those animals often suffer from lack of mental stimulation, insufficient exercise, and inappropriate diets.

Furthermore, the provenance of animals in zoo displays, especially primates, can be of concern. Often primate survivors of medical testing, abandoned after research projects lose funding, are sold to zoos for profit instead of being retired to sanctuaries on a voluntary basis. Some zoos allow breeding of their animals in an effort to perpetuate species that are threatened in the wild. As important as saving threatened species is, and as adorable as a baby gorilla or chimpanzee might seem, seeing one born into a life destined to be lived behind glass, among artificial trees, and facing mimicry and mocking from an often unsympathetic human audience does not seem worth celebrating or even condoning. Although only reputable zoos are accredited members of national organizations such as the Association of Zoos and Aquariums, roadside attractions are permitted to call themselves zoos. The public is often uninformed that such ersatz zoos may be little more than money-making machines for their owners, populated with animals purchased from illicit exotic animal breeders or from illegal hunting in other countries.

Even if a zoo gets its animals from a reputable source, unexpected captive births or fluctuations in funding or public interest sometimes force a zoo to cull its animals from time to time. Although there are some restrictions on how to disperse such animals, including which types of individuals or organizations may buy them, the tracking system used by the Association of Zoos and Aquariums is overwhelmed and fundamentally flawed, making it easy to launder animals. The system is riddled with fraudulent records, whereby the changing of an animal's sex or birthplace creates an easy way to trade in wildlife without fear of leaving behind a paper trail of provenance.

Discarded zoo animals may be sold to exotic animal breeders and handlers, the very same people who may help to stock phony hunts, unscrupulous roadside zoos, the lucrative game-meat industry, and the illegal pet trade. While some may label this practice a quasi-economical use of surplus stock, there are overriding concerns about the life and well-being of each animal. The poor transportation, the conduct of inexperienced or unprepared handlers and owners, and the prospect of a lifetime spent in substandard care so that a marginal entertainment operation can make a few bucks is arguably not an uplifting or morally defensible vision.

The United States Department of Agriculture, which licenses sanctuaries, zoos, and smaller petting zoos, is concerned with violations of the Animal Welfare Act, which relates to living conditions of animals on hand and not necessarily the origins of said animals, whereas the U.S. Fish and Wildlife Service is concerned with the Endangered Species Act and international trafficking. This division still leaves a dearth of enforcement when it comes to interstate trade of exotic species in the United States. Alan Green, author of *Animal Underworld*, an investigation of the exotic animal trade, described it this way:

> Captive-bred wildlife is by and large no one's responsibility, no one's jurisdiction, and really no

one's concern. As a result, the animals are part of an elaborate and sinister shell game, quietly shunted from place to place by those more interested in profits than protection of the species. But search through enough records and follow enough trails, piece together enough evidence from a cross-country dragnet, and much of the deception is ultimately revealed: You can uncover the laundering schemes, pinpoint the animals' destinations, and document how the self-appointed guardians of exotic species are quietly in league with the most disreputable traffickers. At the same time, you can understand how the federal, state, and local laws designed to protect wildlife are flawed, riddled with enough loopholes to permit those bent on exploiting the system to do so with virtual impunity.[630]

After doing some investigative work and trying to locate the origination files of certain exotic animals, Green came to the realization that these animals were truly lost in the shuffle. In fact, the murky paperwork on exotic animals is beneficial to both the seller and the buyer—zoos don't want to be linked to the illegal purchase or sale of a valuable animal, and individual traders don't want to be linked to an exotic animal that they don't have the permits to keep.

A study of studbooks and zoo records by Dr. Stephanie Ostroswki of the Centers for Disease Control and Prevention found that from1975 to 1995, more than 2,500 primates were handed off from zoos to unaccredited institutions.[631] As Alan Green relates, "State veterinarians are interested in cattle and other livestock...State game commissions focus their meager resources on the protection of native animals. The U.S. Fish and Wildlife Service monitors the importation of endangered species but assigns a low priority to captive-bred animals. The USDA, which enforces the federal Animal Welfare Act, has nearly seventeen pages of regulations pertaining to the handling and transportation of dogs and cats, but the care of snow leopards and other wild animals is dismissed in just seven pages."[632]

It's not just in the United States that the animal trade is questionable; there are organizations in Africa, Asia, and Europe skilled at preparing false paperwork and creating lengthy trails to mask the import of illegal or endangered animals into more lucrative markets. This appears to be a case of too little funding and too many other human-centered priorities to make the exotic animal trade worth investigating for most government agencies.

Zoos have a few options when it comes to quenching the public's thirst for exotic animals and their babies, and running what is essentially a profit-based business. The law of supply and demand dictates that zoos provide infant animals for display, which some say also helps to prolong the existence of endangered species, whose numbers are already limited or dwindling. Once the babies grow old enough that they require more space than the zoo is capable of providing, they can be sold either at auctions (where individuals may purchase them) or for laboratory use. It has been proposed that surplus animals be euthanized so they can avoid being lost in a system where animal mistreatment is all too common, but this is clearly a controversial option that is unlikely to gain public approval on account of sanctity-of-life issues. Selective breeding and the promise of lifetime care have also been proposed but are unlikely to be adopted due to financial constraints, as well as consideration of the financial boon that a young exotic animal brings to zoos.

Knowing the history of our relationship with other primates, acknowledging the pros and cons of how we currently use primates, and asking ourselves how we truly feel about the treatment of these close relatives forces us to make decisions about how to proceed in the future. Decisions about how to treat beings that are closely related to humans, yet not quite the same as humans, are not simple, or they would not have generated the amount of discussion and disagreement they do to this day. Clearly such conversations are fraught with implications and seemingly minor details that, combined, would affect not only individual human beings but our culture as a whole.

The fight for rights is often born out of sympathy for the underdog. People are more likely to want to protect animals that have peaceful natures and do not have negative connotations attached in modern culture. In this regard, snakes, vultures, and rats are unlikely to instill the same passion regarding protections as dogs, deer, and, rabbits. Although deer and rabbits may be the scourge of many a suburban garden, these animals do not appear malicious in their intent; conversely, even if testing showed that snakes were more self-aware and cognizant than deer, people would be loath to empathize with a being that slithers on the ground and often plays such a negative role in cultural and religious references.

This discrepancy indicates how subjective the hierarchy we've assigned to the animal kingdom really is. It depends on whims that often grew out of cultures we developed thousands of years ago. As lawyer Steven M. Wise explains his theory about Aristotle's Axiom, in which the Greek philosopher proclaimed Greek man to be the epitome of worth, "No one ever, *ever,* assigns a group to which he or she belongs to any place in a hierarchy of rights other than the top.[633]" The hierarchy of needs we've developed conveniently considers human needs paramount, and our lives have revolved around this flawed assumption.

Humans are loath to change anything to which they've become accustomed, especially if the proposed changes may deny them something that they feel they deserve, whether it be unlimited resources for medical testing, cute, clowning animals at the circus, or fun subjects to study or enjoy at the zoo. As philosopher Bernard Rollin puts it, "So long as powerful vested interests oppose the change, it can become enmired indefinitely, unless public opinion can be galvanized on its behalf."[634] And as primatologist Anne Russon explained to author Shawn Thompson in his book *The Intimate Ape,* "What it would mean is backing off on the notion that we always get to make the decisions and accepting their making the decisions on some things. We want the same land. We want the same foods. We like all the same things, so we're in direct competition in so many areas that it's hard to find a way to live together. We have to accept that the decision should sometimes be in their favor and that means that humans will lose."[635]

The wise philosopher and psychologist R.I.M. Dunbar commented that "the biological reality is that all classifications are artificial. They force a certain order on to the rather chaotic mess of the natural world."[636] Yet, despite the frustrations and the apparent inability to prove anything concrete in terms of humanity's relationship with other primates, we try. We try because we are curious, we try because we think we can prove one thing or another, and we try because we see a bit of ourselves in these animals. We may share approximately 99 percent of our genetic material with certain great apes, but that one percent of difference has proven monumental. The one percent that permits human beings to communicate in complex languages, develop technology capable of putting men on the moon, and discuss such abstract concepts as morality, art, and the purpose of life is such a small physical part of us but a large difference in terms of what it means to be human.

The concept of cladistics involves classifying animals based on anatomical similarities, which leads to inferences about evolutionary proximity. We've seen the problems that anthropomorphic assumptions can lead to, but simply because we are different from the great apes in some ways doesn't mean we should be classified or ethically considered as entirely separate from them.

Anthropologist Michael Ghiglieri claims that if humans were truly classified by biologists with the same level of detachment as other species are labeled, chimpanzees and humans would belong to the same genus (along with bonobos, of course).[637] Paleontologist Steven Jay Gould explains that "[h]umans arise within the space of the *Pongidae*, and cannot represent a separate family, lest we commit the genealogical absurdity of uniting two more distant forms (chimps and gorillas) in the same family and excluding a third creature (humans) more closely related to one of the two united species. I surely cannot claim to be more closely related to my uncle than to my brother, but we make exactly such a statement when we argue that chimps are closer to gorillas than to humans."[638] Proponents of this cladistical change support the inclusion of chimpanzees and bonobos in the *Homo* genus, thus eliminating the need for the genus *Pan*. After all, orangutans, chimpanzees, and gorillas have many differences among them, but surely nobody is debating that we consider them all apes. Why, then, do the differences become paramount when humans are thrown into the biological mix?

Perhaps what is most importance is the idea of personhood. If it is conceded that a being must be a person in order to be deserving of certain rights, we must determine what defines a person.

What does it take to be granted the status of person? If one takes a conservative view, personhood may be defined as having the qualities of self-awareness, introspection, planning for the future, and the ability to ponder abstract concepts and events that are not immediately grounded in the present. If this same conservative view is then carried into the field of ethics, it may be construed that the moral considerations of persons are placed above those of non-persons. After all, due to their increased mental capabilities, persons would ostensibly have the potential for greater suffering when denied certain rights, whereas the base cognition of non-persons would mean that they wouldn't even notice or care about a lack of general rights. (Of course, when applying this concept to an entire population, it's generally accepted that the average intelligence would be considered, rather than evaluating the mental capabilities of individuals. This method, then, would not count the deficiencies of the mentally handicapped towards an evaluation of the intellectual capabilities of the whole.)

If, however, a more liberal view is applied to this situation, it's possible to consider a person as any being capable of appreciating or enjoying the rights in question. This would circumvent the argument that granting certain legal rights to nonhuman primates could assume, for example, the right to vote, because nonhuman primates would not appreciate, enjoy, or have any intention to vote in an election. They would, however, certainly appreciate the right to live free from pain and imprisonment, in a manner natural to them.

It makes sense to grant rights to those that could enjoy or appreciate the granting of such rights. Some nonhuman beings have the need to fly in open spaces, graze on fresh grass, or make beds in certain types of trees in order to be healthy and contented. Human beings might not appreciate the freedom to do these specific things, although there are other things, such as religious freedom, that humans deem very important. It's important to realize that rights don't need to be universal or identical within the moral community, only that they be equal in scope.

Another view of personhood involves abilities of consciousness. Rationality, and the acknowledgement that others have similar rational mentalities replete with desires and emotions such as one's own, reveal the personhood of the agent. Accepting the ego states of others defines the higher thinking of a person, according to philosopher and cognitive scientist D.C. Dennett. Dennett continues on to say that the possible person in question must also be able to communicate verbally the observation of the intentions of others[639], but it must be considered that some beings, including some intelligent, aware humans, are unable to communicate verbally, although their status of personhood should not come into question.

Since nonhuman primates are unable to communicate verbally because of physical limitations, this should not stand in the way of consideration of personhood, arguably in much the same way as a mute human being (or, indeed, one suffering paralysis, for instance) should be considered. After all, as sign language studies have shown, nonhuman primates are certainly able to communicate via signs (including making observations about the mental states of others), as do humans who are deaf, for example, and this would further seem to prove that nonhuman primates are worthy of consideration as persons. As has been mentioned, studies of primates in the wild have shown that their natural, native utterances and calls can indicate empathy, sympathy, and even intentional deception, which seems to prove their awareness of the mental states of others.

It may be interesting to note that anytime the personhood of populations of humans has come into question throughout recorded history—whether those population might be females, tribal groups, ethnic groups, or the disabled—the characteristics discussed as markers of personhood have all been proven to be present in nonhuman primates. Mental abilities such as the comprehension of temporality, consciousness of others, and curiosity about change are just a few of the characteristics under scrutiny over time, and any sampling of primate studies will show that these mental states are often shown to be exhibited by primates, both in the wild and in captivity.

Although at times the denial of full rights has been detrimental to those deemed by society as of a lower class or lacking in personhood, at times extra rights can actually be granted to a being of lesser power. Consider, for example, child labor laws or Title X rulings in school sports, both of which were created to protect groups that could not protect themselves from either being taken advantage of or discriminated against.

As Harlan B. Miller points out in his essay "The Wahokies," although these rulings were passed in order to protect those not fully capable of protecting themselves, the rulings themselves are essentially paternalistic, and their very existence seems to imply that those in power (the majority) know what is best for all (including the minority in question).[640] As history has shown, this is not always the case. While it may not be better to discriminate against or purposely harm a given population, it also seems somewhat aggressive and presumptive to assume to know what is best for a population with needs other than the ruling majority's.

Philosopher Stephen Clark has found a way to use both sides of the "what defines humanity" debate to the advantage of nonhuman primates. In the following excerpt, he responds to those who would say that humans are superior to others in the animal kingdom because of having been created in God's image: "If we are apes, let us be apes together. If we are 'apes' [as in aping the Divine], let us acknowledge what our duty is as would-be saints and give the courtesy we owe to those among whom we sprang. Either we evolved along with them, by the processes described elsewhere, or else we evolved, in part, to imitate a Divine Humanity. Neither theory licenses a radical disjunction between us and other apes. Either may give us reason to esteem and serve the greater humankind."[641] Since our biology is shared with apes, inclusion in the realm of the humankind that Clark speaks of would entail protections for nonhuman primates.

Searching for ways to incorporate interspecies harmony within the dogma of established religions is one way to integrate increased legal protections for nonhuman primates into the existing belief systems of much of the world. An alternate consideration is the evolutionary worldview, which asserts that humans are a species living in tandem with many other species on the planet. The evolutionary worldview's absence of any hierarchy helps to democratize the plight of nonhuman primates. Additionally, we can postulate that the dominion that some people believe is granted by God to humans over animals may actually be one of stewardship instead of dominance. Perhaps man was granted superior reasoning skills and language properties so as best to care for the other creatures sharing the planet. Much in the way that humans give more attention and care to disabled people that are unable to care for themselves, perhaps members of the animal kingdom who are helpless in other ways deserve even more protections than humans enjoy. Perhaps, rather than ignoring their potential legal rights, we should be investigating those rights.

James Rachels, a philosophy instructor at the University of Alabama, invokes the teachings of Aristotle to imply that like cases should be treated alike. "I take this to mean that individuals are to be treated in the same way *unless there is a relevant difference between them*...[but] where there are no relevant differences, they must be treated alike."[642] In other words, it would not be necessary to treat all animals the same as humans, for there are distinct differences between some animals and humans that would make similar treatment absurd.

It is not necessary for two populations to be almost identical in order to receive equal treatment; for instance, women and men are different in many ways, but it is universally accepted that the two groups are deserving of equal treatment in many respects. This should not be taken to mean that all treatment enjoyed by one population ought to be assigned to the other, similar population, for it may not benefit them, or they may not want or need to partake of that particular treatment. It may not be in their best interest. Professors Heta and Matti Häyry refer to this concept as "equality of consideration."[643] This idea leads to the concept that liberties might be granted to beings depending on the varying depth of their abilities.

For example, they go on to say, "There are animals, including great apes like ourselves, who, in addition to their ability to suffer from constraint and physical pain, are aware of themselves as distinct entities whose existence is temporally continuous. Only these individuals can suffer from their own demise, or the thought of their own nonbeing in the future, and only they possess the right to life in the strict sense. There are also animals who lack self-awareness but who are sentient and capable of being distressed if they are imprisoned—they have the rights to liberty and the avoidance of torture. Finally, beings who are merely sentient, i.e., only sensitive to physical suffering, are entitled to protection against deliberately inflicted pain."[644]

A meritocracy is a community where rights are given based on the utility of each individual, where everyone gets their just desserts, so to speak. This argument is not well applied when it comes to primates, because not only would it mean that many humans would not enjoy equal liberties (after all, how many people do you know who directly worked to create the system of protections that govern our societies?), but it would mean that many primates would deserve more protections than humans. The chimpanzees that donated decades of their lives to medical research surely should be able to gain some freedom and rights from their years of service.

As moral agents, most humans are capable of making choices and informed decisions about the way they handle things in their world. As moral patients, animals and other humans who are not able to make informed decisions, such as infants and mentally disabled people, are in a submissive stance when it comes to their treatment. Modern culture tends to grant power and responsibility to moral agents, evident in the punishments and disapproval our culture gives to abusers of animals and children. At some point, though, a line is drawn in the consideration of morality as it relates to animals. People who believe that animals are happy living in captivity because they have their basic needs met—be it as pets, actors, or in zoos or labs—would certainly resist imprisonment, themselves. This is because the psychological component is powerful, and it is fairly obvious that the basic needs for survival are but a small part of what we all would want to experience in our lives.

People may say that primates shouldn't be afforded more rights because they don't have the ability to fight and gain those rights themselves. While it's true that primates don't have the language and organizational skills to band together and protest against humans for equal rights, some groups of human beings also do not have that ability, such as the developmentally or physically disabled. On their behalf, able-bodied people have acted to gain freedoms and protections for them. This is no different than human beings doing this for nonhuman primates. As the Häyrys put it, "Protection and compassion cannot with good conscience be restricted to those who know how to earn them."[645] Yet, the autonomy granted to severely disabled humans is denied other primates, including those with greater mental acuity than some of the disabled humans.

Tom Regan, a philosophy professor at North Carolina State University, believes that people should rely on beings' inherent worth to answer questions regarding their freedoms. The fact that nonhuman primates lack the ability to converse with us in a spoken language, or the fact that their intellectual abilities are not as strong as the average adult human being, does not diminish their inherent worth. Regan believes that animals have lives that matter to them, which illustrates their inherent self-worth. It is morally indefensible, he believes, to use a being with inherent self-worth in a way that is not in their own best interest. To use them to benefit others is to act in direct opposition to their individual welfare. Any inflicted harm or deprived freedom (which is inherently harmful to a being's welfare) is cruel and not to be tolerated.

The concept of rights is based in a teleological view of the world. The telos of a being is its intrinsic nature, or its purpose (or ultimate goal). Laws exist to protect a human individual's right to life, liberty, and the pursuit of happiness, but what of the worth of other creatures? Although the telos of a chimpanzee may be different than the telos of a human being, the difference does not imply that it is any less worthy of protections. Once again, *different* does not automatically mean *less than*. Why withhold rights from nonhuman primates? Should humans be required to withhold rights from other species until all humans can enjoy satisfactory rights? This would seem to imply that humans are more worthy than other species, but should the significance of humans be considered greater than that of other species?[646]

Philosopher Bernard Rollin states, "One major step towards extending the ethic to animals, not difficult for the average person to take, is the realization that there exists no good reason for withholding it: in other words, that there is no morally relevant difference between humans and animals which can rationally justify not assessing the treatment of animals by the machinery of our consensus ethic for humans. Not only are there no morally relevant differences, there are significant morally relevant similarities."[647]

Animal rights lawyer and advocate Gary Francione has written extensively on the topic and believes that nonhuman primates, along with other animals, deserve rights solely because they are sentient. In a forward to one of his books, the Department of Philosophy at Bucknell University summed up Francione's ideology this way: "Specifically, he argues that all sentient beings, those capable of experiencing pleasure and pain, have a fundamental interest in avoiding suffering and continuing to exist...[that] all humans have the right not to be treated as the property of others. Francione contends that there is no reason not to accord this right to nonhuman animals as well."[648] Francione rejects any notion of anthropocentrism, or basing rights in connection to a species' relatedness to humans, and believes that sentiency is itself proof of a being's desire for a life unencumbered by service or suffering. He believes that sentiency, combined with their intelligence and awareness, warrants bestowing legal personhood on great apes.

Professor Joseph Raz has stated that, under his reciprocity thesis of the interest theory of rights, animals should not be afforded rights, because animals are not members of the same moral community as humans and, thus, cannot possess rights under our moral code.[649] One response to Raz's theory is that it is entirely possible for animals to exist within humanity's moral sphere, and that it is only outdated historical thought that places them outside this area of consideration. The argument only gains strength when one reviews the biological relationship between nonhuman and human primates.

Of course, even if the rather large hurdle of agreement that personhood (or some similar legal status of equality) should be granted to nonhuman primates is overcome, the question of morality remains. Human beings are expected to adhere to codes of morality, which can certainly vary among populations, although there is a general expectation to respect the lives of others and tread lightly on their freedoms. Should nonhuman primates be expected not only to understand but also adhere to this often shifting moral code? If this seems excessive, then does it seem hypocritical to grant them entrance into a community whose rules they may not be capable of following fully? When there is a conflict, what should be the procedure, considering that many nonhuman primates are easily capable of physically overpowering any human? Would there be interspecies responsibility or accountability? It's already illegal for humans to kill most nonhuman primates, but if a nonhuman kills a human, would there be punishment for murder or manslaughter?

The rules of acceptable behavior in primate communities can vary greatly from those in human communities, so is it fair to decide that nonhumans must not injure humans, simply because we have the language to state the case? The fact remains that all day long, every day, primates could loudly be declaring the opposite, that humans may not injure nonhumans, but because of a language barrier and our own disinterest, we would not hear a thing.

Animal activists are a varied population that runs the gamut from peaceful educators to groups considered terrorists by the federal government. Most animal rights groups place the interests of nonhuman primates high on their list of concerns, due both to primates' high level of intelligence and to the uses and abuses of nonhuman primates in human societies. While those who support animal rights may be referred to as animal welfarists, animal advocates, protectionists, zoophiles, humanitarians, rightists, and liberationists,[650] what should those who specifically support primate rights be called? The word primatologist is a scholarly and vocational title, not an indication of ethical or moral judgments or concerns. The term primate rightists is cumbersome, at best, and potentially confusing or mystifying, at worst.

There is also the distinction between proponents of animal rights (who question any differences in protections based on species membership) and animal welfare (who take a stance against cruelty to animals but not their practical uses for human consumption, protection, decoration, etc). As outlined in the first chapter of this book, animal welfarists subscribe to the hierarchical placement of humans over nonhumans and see nothing wrong with humans asserting their dominance by using animals to benefit their own or even other species. To apply this distinction to the world of nonhuman primates, a rightist might wish to end the nonhuman primate pet trade, for instance, while a welfarist might simply support the proper care of capuchins that assist disabled people.

Some people believe that the more calm and controlled an activism event is, the more likely that positive change will come about as a result. Others believe that any news bringing headlines and attention to the cause for animal rights is a good thing, even if it includes group member arrests for trespassing, threatening, or even acts of terrorism. Shock can certainly go a long way toward searing an image into someone's brain, and anyone who's ever done an internet search on vivisection, or even just on animal testing, can verify that the results are nothing short of horrific to view. As with any human rights issues, if posting such disturbingly violent and macabre images is capable of changing one person's mind, can it be said that it's extreme or overdramatic, particularly if it is accurate? If changing minds is the goal, and these methods save or improves the lives of animals living in captivity, extreme action may not be so easy to condemn at second glance.

If one agrees with any pyramidal ranking of animals, it seems clear that if humans are on top (which seems to be universal among cross-species hierarchies), nonhuman primates are one step below in the great ranking. Specifically, the nonhuman great apes (the rest of the family Hominidae, which includes bonobos, gorillas, chimpanzees, and orangutans, in addition to humans) are a very small step below human beings in the Linnaean classification system, as well as genetically. If genetic proximity is a factor, then great apes should logically be the next disenfranchised group to receive increased legal protection and recognition of appropriate rights (those that have value to them), including the right to live a life free of pain, mistreatment, and abuse.

If humans can determine how best to handle increased legal rights for great apes, the door might open to further change that would result in other nonhuman primates being granted increased legal protection, as well. The transition would not be seamless, and accommodations would surely need to be made to adjust certain systems to the new changes, but this has occurred in the past when legal rights were bestowed upon oppressed groups. The biggest hurdle is mental, as humans tend to fear the unknown and often cannot imagine a future different from their present circumstances.

Even if humans decided to commit more seriously to rights for nonhuman primates, that would not mean the release of all currently captive-held primates to run free throughout the world. Indeed, that would not be in their best interest. The most compassionate response would be to allow the animals already in captivity to live out their lives in comfort and safety, with as much ability to satisfy natural behaviors as possible, and breeding and exporting of captive animals would be halted. We would then need to recommit to protecting their indigenous environments. In a few decades all primates living in captivity would have died, and the remaining nonhuman primates would have spent their entire lives in the wild. Eventually, it's possible that the entire order of *Primates* would be granted equal opportunity to direct their own lives in the habitat they desire, be it orangutans in the wilds of Borneo or perhaps baboons on the plains of Texas.

Closing

And I am my brother's keeper,
And I will fight his fight,
And speak the word for beast and bird,
Till the world shall set things right.[651]
- Ella Wheeler Wilcox

In the not so distant past history of human beings, the idea that animals should be granted legal rights was once viewed as ridiculous, even obscene. The idea was used as a parallel to emphasize the absurdity of women obtaining legal rights.

Many people now believe that sentient beings should have rights to things they can recognize as significantly affecting their lives. Nonhuman primates might not be able to comprehend voting, for example, but freedom from the infliction of pain? Surely every living species would enjoy this right.

Charles Darwin believed that humanity's sense of morality evolved over time through natural selection. Darwinism claims that at first, humans only include those who are immediately like them into their sphere of moral consideration, but over time this sphere can be widened (following the discovery of beings that are deemed worthy of care and attention) to eventually encompass other races, other genders, and even other species.[652]

Darwin wrote, "Looking to future generations, there is no cause to fear that the social instincts will grow weaker, and we may expect that virtuous habits will grow stronger, becoming perhaps fixed by inheritance. In this case the struggle between our higher and lower impulses will be less severe, and virtue will be triumphant."[653]

An ethical consideration of the interests of nonhuman primates is warranted. The concept that other primates' brains are less complex and that those primates are consequently less intelligent than humans ought not be used to justify their subservience to the human race. Even intellectual might does not automatically make right.

Speciesism is no less cruel than racism and sexism. As American founding father Thomas Jefferson eloquently expressed in an argument against slavery, "Because Sir Isaac Newton was superior to others in understanding, he was not therefore lord of the property or persons of others."[654] Abolitionist Sojourner Truth made a similar comparison about basing one's rights on intellectual capabilities, whether actual or hypothesized. "What's that got to do with women's rights or Negroes' rights? If my cup won't hold but a pint and yours holds a quart, wouldn't you be mean not to let me have my little half-measure full?"[655]

There are many correlations between human slavery and the ways that primates are currently treated and used. What ended human slavery? As industrial development and other elements of progress relieved man from various tasks, there was a simultaneous rising awareness of human rights and the blind injustice of racism. What will end the exploitation of nonhuman primates? Creativity and technological developments such as computer-generated imagery that no longer necessitate using live performers to feed the public's hunger for primate entertainers? Will it also include a higher level of thought, similar to that which occurred before the Civil War, demanding freedom for those living under the control of others? Ending speciesism has proven to be just as ugly a fight as ending racism, if not more difficult, because the minority in question is even more an underdog than ever before—without a voice, without any historical legs to stand on, and without the ability to rise up and organize its own revolt.

A major shift in thinking often begins with one seemingly minor action committed by one person. This somehow catches people's attention because it contradicts the norm. It's typically a novel idea that becomes a widespread topic of discussion. Once the matter is discussed enough, the new view no longer seems so strange, and people slowly begin to see that there may be some logic behind it, after all. Some brave souls eventually begin to adapt their lives to include the newfound truth, believing that any sacrifices they may endure are warranted and justified by the information now incorporated into their lives.

As the field of primatology grows in scope and depth, more and more conclusions are drawn showing that human beings share many characteristics with nonhuman primates. There has been a shift in the definition of *human*. Perhaps those qualities that we had thought made us unique are not solely ours, after all, but traits that we share with other species. Tool use, problem-solving capacity, and even language skills have been exhibited by nonhuman primates, sometimes in ways that exceed some human beings' abilities.

Change is hard but is sometimes inevitable. As the boundary between human and nonhuman primates grows increasingly thinner, it becomes more and more difficult to substantiate our assumed superiority over others in the animal kingdom.

"I have often wondered what it would be like to suddenly discover that you are not who you thought you were," wrote primatologist Roger Fouts about Washoe the chimpanzee adjusting from living like a human to living the natural life of a member of her own species. "Would we be like Washoe and accept it and show compassion and caring for our newly discovered conspecifics? Or would we maintain our earlier arrogance and continue to oppress and refuse to accept our own kind?"[656]

Washoe chose the high road.

References

Adams, Chris. "Some Chimps Never Recover From Stresses of Research".
    McClatchy. April 24, 2011.
    http://www.mcclatchydc.com/2011/04/24/112432/some-chimps-never-
    recover-from.html.

Altmann, Jeanne. "Introductory Remarks." Introduction to the Scientists
    Center for Animal Welfare's conference Well Being of Nonhuman
    Primates in Research, Bethesda, Maryland, June 23, 1989, ed. Joy A.
    Mench and Lee Krulisch, 21-22.
    http://www.scaw.com/assets/files/1/files/nhp-.pdf.

American Humane Association. "Red Star Animal Emergency Services
    History." Accessed 2009.
    http://www.americanhumane.org/protecting-
    animals/programs/animal-emergency-services/history.html.

American Humane Association. "Who We Are." Accessed 2013.
    http://www.americanhumane.org/about-us/who-we-are.

American Sanctuary Association. "About American Sanctuary Association
    (ASA)." Accessed 2008.
    http://www.asaanimalsanctuaries.org/about_ASA.htm.

American Sanctuary Association. "Animal Sanctuaries." Accessed 2008.
    http://www.asaanimalsanctuaries.org.

Anderson, Stephen R. *Doctor Doolittle's Delusion: Animals and the
    Uniqueness of Human Language*. New Haven: Yale University Press,
    2006.

Animal Legal & Historical Center, Michigan University College of Law.
    "ALDF v. Glickman (standing)." Accessed 2010.
    http://www.animallaw.info/cases/caus154f3d426.htm.

Animal Legal & Historical Center, Michigan University College of Law.
    "Chimpanzee Health Improvement Maintenance and Protection
    Act." Accessed 2010.
    http://www.animallaw.info/statutes/stusfdpl106_551.htm.

Animal Legal & Historical Center, Michigan University College of Law. "CITES." Accessed 2009. http://www.animallaw.info/treaties/itcites.htm.

Animal Legal & Historical Center, Michigan State University College of Law. "US Animal Welfare Act Regulations Subpart D, Primates." Accessed 2013. http://www.animallaw.info/administrative/adusfdawaregd.htm.

Animal Legal & Historical Center, Michigan State University College of Law. "US Chimpanzee Sanctuary." Accessed 2010. http://www.animallaw.info/statutes/stusfd42usc287a_3a.htm.

Animal Legal Defense Fund. "About Us." Accessed 2009. http://aldf.org/article.php?list=type&type=3.

Animal Legal Defense Fund. "A New Life For Angel, Cody, and Sable!" Accessed 2007. http://www.aldf.org/article.php?id=369.

Animal Legal Defense Fund. "Animal Law Program." Accessed 2009. http://aldf.org/article.php?list=type&type=15.

Animal Legal Defense Fund. "Animal Legal Defense Fund." 2009. YouTube. Posted January 10, 2008. http://www.youtube.com/watch?v=0tPjVqDiB_I&feature=player_embedded.

Animal Legal Defense Fund. "Landmarks and Victories." Accessed 2009. http://www.aldf.org/article.php?list=type&type=81.

Animal Legal Defense Fund. "Student Animal Legal Defense Fund Chapters." Accessed 2009. http://aldf.org/article.php?id=446.

Animal Rights History. "Animal Rights Law." Accessed 2009. http://www.animalrightshistory.org/index.htm.

Animal Rights History. "Cruel and Improper Treatment of Cattle Act 1822." Accessed 2013. http://www.animalrightshistory.org/animal-rights-law/romantic-legislation/1822-uk-act-ill-treatment-cattle.htm.

Animal Rights History. "Cruelty to Animals / Anti-Vivisection Act 1876; Great Britain Parliament." Accessed 2011. http://www.animalrightshistory.org/animal-rights-law/victorian-legislation/1876-uk-act-vivisection.htm.

Association of Zoos and Aquariums. "Becoming Accredited." Accessed 2010. http://www.aza.org/becoming-accredited/.

Association of Zoos and Aquariums. "Health, Husbandry, and Welfare." Accessed 2009. http://www.aza.org/health-husbandry-and-welfare/.

Association of Zoos & Aquariums. "List of Accredited Zoos and Aquariums." Accessed 2010. http://www.aza.org/current-accreditation-list/.

Baeckler, Sarah, MS. "Campaign to End the Use of Chimpanzees in Entertainment." October 14 2003. Testimony presented at a briefing co-hosted by the Chimpanzee Collaboratory and the Environmental Media Association, Los Angeles, California, October 14, 2003. http://www.primatepatrol.org/pdfs/undercover_at_a_training_f acility.pdf.

Ballantyne, Coco. "The Lobbying Landscape and Beyond: 15 Groups to Know." *Nature Medicine.* Accessed 2008. http://www.nature.com/nm/journal/v14/n10/full/nm1008-1002.html.

BBC News. "Chimps 'Feel Death Like Humans'." Accessed 2010. http://news.bbc.co.uk/2/hi/science/nature/8645283.htm.

Beers, Diane L. *For The Prevention of Cruelty: The History and Legacy of Animal Rights Activism in the United States.* Athens: Swallow Press / Ohio University Press, 2006.

Bekoff, Marc. *Animals Matter.* Boston: Shambhala Publications, Inc., 2007.

Bekoff, Marc. "Common Sense, Cognitive Ethology and Evolution." In *The Great Ape Project: Equality Beyond Humanity*, edited by Paola Cavalieri and Peter Singer, 102-107. New York: St. Martin's Press, 1993.

Bideawee. "Bideawee History." Accessed 2007. http://www.bideawee.org/about_bideawee/our_organization/history.p hp.

Bideawee. "History." Accessed 2012. http://www.bideawee.org/History.

Blair, Craig J, DVM. "Pets or Prisoners?" *Southsider Magazine,* June 2005. Accessed 2013. http://petmonkeyinfo.org/pets_or_prisoners.htm.

Blum, Debrorah. *The Monkey Wars.* New York: Oxford University Press, 1994.

Borneo Orangutan Survival. "10 Years." Accessed 2009. http://savetheorangutan.org/splash/nm10.pdf.

Borneo Orangutan Survival Australia. "Mawas." Accessed 2012. http://www.orangutans.com.au/Orangutans-Survival-Information/Mawas.aspx.

Borneo Orangutan Survival Australia. "Samboja Lestari." Accessed 2012. http://www.orangutans.com.au/Orangutans-Survival-Information/Samboja-Lestari-128959069771931264.aspx.

Born Free Foundation. "About Zoo Check." Accessed 2009. http://www.bornfree.org.uk/campaigns/zoo-check/about-zoo-check/.

Born Free Foundation. "Born Free's History." Accessed 2009. http://www.bornfree.org.uk/about-us/history/introduction/.

Born Free USA. "API's History." Accessed 2009. http://www.bornfreeusa.org/d_about.php.

Born Free USA. "Born Free USA Primate Sanctuary: A More Natural Life." Accessed 2009. http://www.bornfreeusa.org/a8_sanctuary.php.

Born Free USA. "Born Free USA's History." Accessed 2009. http://www.bornfreeusa.org/d_about.php.

BornFree USA. "Summary of State Laws Relating to Private Possession of Exotic Animals." Accessed 2011. http://www.bornfreeusa.org/b4a2_exotic_animals_summary.php.

Boyd Group. "Paper 2: Empirical Evidence on the Moral Status of Non-Human Primates." Accessed 2013. http://www.boyd-group.demon.co.uk/Paper2.pdf.

Boyd Group. "Paper 3: The Moral Status of Non-human Primates: Are Apes Persons?" Accessed 2013. http://www.boyd-group.demon.co.uk/Paper3.pdf.

Britten, R.J. "Divergence Between Samples of Chimpanzee and Human DNA Sequences is 5% Counting Indels." *Proceedings National Academy Science*, October 15 2002. Accessed 2013. http://www.pnas.org/content/99/21/13633.long.

Brubaker, Bill. "Mission Orangutan." *Smithsonian.* December 2010, 36-45.

Butler, Rhett A. "Why We Are Failing Orangutans." *Mongabay.com,* March 1, 2010.   Accessed 2013. http://news.mongabay.com/2010/0301thompson_orangutans_in terview.html.

Cantor, David. "Items of Property." In *The Great Ape Project: Equality Beyond   Humanity*, edited by Paola Cavalieri and Peter Singer, 280-290. New York: St. Martin's Press, 1993.

Cantrell, Cindy. "Monkey's Business is Lending a Hand." *Boston Globe*, December 5, 2010. Accessed 2013. http://www.boston.com/news/health/articles/2010/12/05/ monkey_becomes_concord_mans_assistant_after_car_accident /.

Capital News Service. "Deficit Hawk Swoops to Save Chimps from U.S. Research". *The Seattle Times,* April 13 2011. Accessed 2013. http://seattletimes.nwsource.com/html/health/2014767479_chi mps14.html.

Captive Wild Animal Protection Campaign. "About CWAPC." Accessed 2010. http://www.cwapc.org/about/index.html.

Carbone, Larry. *What Animals Want: Expertise and Advocacy In Laboratory Animal Welfare Policy.* New York: Oxford   University Press, 2004. Accessed 2013. http://books.google.com/books?id=Iheg3hkj99AC&printsec=fr ontcover#v=onepage&q=&f=false.

Catalan, Thomas. "Apes Get Legal Rights in Spain, to Surprise of Bullfight Critics." *The Times.* June 27, 2008. Accessed 2012. http://www.timesonline.co.uk/tol/news/world/europe/article4 220884.ece.

Cavalieri, Paola and Peter Singer, ed. *The Great Ape Project: Equality Beyond Humanity.* New York: St. Martin's Press, 1993.

Center for Great Apes. "We Feed The CareerBuilder Chimpanzees." Accessed 2011. http://www.centerforgreatapes.org/news.aspx#We%20Feed%20the%20CB%20Chimpanzees.

Centers for Disease Control and Prevention. "B Virus (herpes B, monkey B virus, herpesvirus simiae, and herpesvirus B)." Accessed 2010. http://www.cdc.gov/herpesbvirus/.

Channel3000. "Workers Discuss Ethics of Primate Research." Accessed 2010. http://www.channel3000.com/technology/23591299/detail.htm.

Chantek Foundation Historical Site. "Project Chantek." Accessed 2013. http://chantek.org/project-chantek.

Chimpanzee Sanctuary Northwest. "Missy." Accessed 2012. http://www.chimpsanctuarynw.org/the_chimpanzees/chimp/missy.

Chimp Haven. "Our History." Accessed 2013. http://www.chimphaven.org/about-us/our-history/.

CITES. "Mammals." Accessed 2013. http://www.cites.org/gallery/species/mammal/mammals.html.

Clark, Stephen R.L. "Apes and the Idea of Kindred." In *The Great Ape Project: Equality Beyond Humanity*, edited by Paola Cavalieri and Peter Singer, 113-125. New York: St. Martin's Press, 1993.

Cohen, Jon. "Thinking Like A Chimpanzee." *Smithsonian*. September 2010: 50-57.

Conn, P. Michael. "A Legal Challenge To Animal Research." *TheScientist.com,* January 1, 2009. Accessed 2013. http://www.the-scientist.com/article/display/56167.

Convention on International Trade of Endangered Species of Wild Flora and Fauna. "Convention on International Trade of Endangered Species ofWild Flora and Fauna:Appendices I, II, and III." Accessed 2013. http://www.cites.org/eng/app/appendices.php.

Convention on International Trade in Endangered Species of Wild Fauna and Flora. "Text of the Convention." Accessed 2013. http://www.cites.org/eng/disc/text.php.

Corbey, Raymond. "Ambiguous Apes." In *The Great Ape Project:*
Equality Beyond Humanity, edited by Paola Cavalieri and Peter
Singer, 126-136. New York: St. Martin's Press, 1993.

Correll, John T. "The Astrochimps." *Air Force Magazine,* September   2011.
Accessed 2013.
http://www.airforcemag.com/MagazineArchive/Pages/2011/Se
ptember%202011/0911astrochimps.aspx.

Cosmides, Leda and John Tooby. "Evolutionary Psychology: A Primer.
Center for Evolutionary Psychology." University of California,
Santa Barbara. Accessed 2013.
http://www.psych.ucsb.edu/research/cep/primer.html.

Darwin, Charles. *The Descent of Man and Selection in Relation to Sex.* New
York: D.Appleton and Company, 1871.

Department of the Interior, Fish and Wildlife Service. "Endangered and
Threatened Wildlife and Plants; 90-Day Finding on a Petition to List
All Chimpanzees (Pan troglodytes) as Endangered." *Federal*
Register, September 1, 2011. Accessed 2013.
http://www.fws.gov/policy/library/2011/2011-22372.pdf.

De Waal, Frans and Frans Lanting. *Bonobo: The Forgotten Ape.*
Berkeley and Los Angeles: University of California Press, 1997.

De Waal, Frans. "Empathy in Primates and Other Mammals." In
Empathy: from Bench to Bedside, edited by J. Decety, 87-106.
Cambridge: MIT Press, 2011.
http://www.emory.edu/LIVING_LINKS/publications/articles/d
eWaal_2011a.pdf.

De Waal, Frans. *My Family Album: Thirty Years of Primate Photography.*
Berkeley and Los Angeles: University of California Press, 2003.

De Waal, Frans. *Primates and Philosophers: How Morality Evolved,* edited by
Stephen Macedo and Josiah Ober. Princeton: Princeton University
Press, 2006.

Diamond, Jared. "The Third Chimpanzee." In *The Great Ape Project:*
Equality Beyond Humanity, edited by Paola Cavalieri and Peter
Singer, 88-109. New York: St. Martin's Press, 1993.

Dolhinow, Phyllis and Augustin Fuentes. *The Nonhuman Primates.*
Mountain View: Mayfield Publishing Company, 1999.

Dunbar, R.I.M. "What's In A Classification?" In *The Great Ape Project:Equality Beyond Humanity*, edited by Paola Cavalieri and Peter Singer, 110-113. New York: St. Martin's Press, 1993.

"Encephalomyelitis." Accessed 2010. Wikipedia. http://en.wikipedia.org/wiki/Encephalomyelitis.

Encyclopaedia Brittanica. "James Burnett, Lord Monboddo." Accessed 2010. http://www.britannica.com/EBchecked/topic/389017/James-Burnett-Lord-Monboddo.

English, Dori. "A Monkey in Your Home?" Accessed 2013. http://petmonkeyinfo.org/monkeyinyourhome.htm.

Estrada, Alejandro. "Human and Non-human Primate Co-existence in the Neotropics: a Preliminary View of Some Agricultural Practices as a Complement for Primate Conservation," *Ecological and Environmental Anthropology*, 2006.

Europa: Summaries of EU Legislation. "Fundamental Rights and Non-Discrimination." Accessed 2013. http://europa.eu/legislation_summaries/institutional_affairs/trea ties/amsterdam_treaty/a10000_en.htm.

European Cetacean Bycatch Campaign. "Animal Welfare and the Treaty of Amsterdam." Accessed 2010. http://www.eurocbc.org/page673.html.

Favre, David. "Overview of U.S. Animal Welfare Act." Last modified May 2002. http://www.animallaw.info/articles/ovusawa.htm.

Favre, David and Vivien Tsang. *The Development of the Anti-Cruelty Laws During the 1800's*. Detroit: Detroit College of Law Review, 1993.

Finkelmeyer, Todd. "Campus Connection: Animal Rights Group Wants Answers from USDA." *The Cap Times*. February 19, 2010. http://host.madison.com/news/local/education/campus_connect ion/campus-connection-animal-rights-group-wants-answers-from-usda/article_5f44b4c4-1ccc-11df-9a72-001cc4c03286.html.

Fossey, Dian. *Gorillas In The Mist.* New York: Houghton Mifflin, 1983.

Fouts, Roger and Stephen Turkel Mills. *Next of Kin: My Conversations with Chimpanzees.* New York: HarperCollins Publishers, Inc, 1997.

Fouts, Roger S. and Deborah H. Fouts. "Chimpanzees' Use Of Sign Language." In *The Great Ape Project: Equality Beyond Humanity*, edited by Paola Cavalieri and Peter Singer, 28-41. New York: St. Martin's Press, 1993.

Francione, Gary L. *Animals as Persons: Essays on the Abolition of Animal Exploitation.* New York: Columbia University Press, 2008.

Francione, Gary L. *Animals, Property, and the Law.* Philadelphia: Temple University Press, 2007.

Francione, Gary L. "Personhood, Property, and Legal Competence." In *The Great Ape Project: Equality Beyond Humanity*, edited by Paola Cavalieri and Peter Singer, 248-257. New York: St. Martin's Press, 1993.

Friends of Washoe. "Tributes to Washoe." Accessed 2013. http://www.friendsofwashoe.org/meet/washoe_memorial.html.

Gardner, James Ross. "Why Is This Chimp Smiling?" *Seattle Met Magazine,* December 2009. http://www.seattlemet.com/news-and-profiles/articles/chimpanzee-sanctuary-northwest-1209.

Global Federation of Animal Sanctuaries. "Global Federation of Animal Sanctuaries." Accessed 2010. http://sanctuaryfederation.org/gfas-sanctuaries.php.

Goodman, Justin. "Animal Tests are Today's Tuskegee Experiments," *The Sacramento Bee,* March 14, 2011. http://www.sacbee.com/2011/03/14/3473727/animal-tests-are-todays-tuskegee.html.

Goodman, Justin. "The Argument Against Laboratory Testing on Animals," *Sign On San Diego,* January 15, 2011. http://www.signonsandiego.com/news/2011/jan/15/the-argument-against-laboratory-testing-on-animals/.

Gorman, James. "Agency Moves to Retire Most Research Chimps," *The New York Times,* January 22, 2013. http://www.nytimes.com/2013/01/23/science/nih-moves-to-retire-most-chimps-uscd-in-rcscarch.html?_r=2&.

Gorman, James. "U.S. Suspends Use of Chimps in New Researach," *The New York Times,* 15 December 2011. Accessed 2013. http://www.nytimes.com/2011/12/16/science/chimps-in-medical-research.html?_r=1.

Govtrack.us. "H.R. 1326: Great Ape Protection Act of 2009." Accessed 2010. http://www.govtrack.us/congress/bill.xpd?bill=h111-1326.

Govtrack.us. "H.R. 5566: Animal Crush Video Prohibition Act of 2010." Accessed 2010. http://www.govtrack.us/congress/bill.xpd?bill=h111-5566.

Govtrack.us. "S.3694: Great Ape Protection Act of 2010." Accessed 2010. http://www.govtrack.us/congress/bill.xpd?bill=s111- 3694.

Great Ape Survival Project: Kinshasa Declaration on Great Apes, United Nations Educational, Scientific and Cultural Organization (September 9, 2005). http://www.unesco.org/mab/doc/grasp/E_KinshasaDeclaration.pdf.

Green, Alan and The Center for Public Integrity. *Animal Underworld: Inside America's Black Market For Rare and Exotic Species.* New York: Public Affairs / Perseus Books Group, 1999.

Grehan, John R. "Mona Lisa Smile: The Morphological Enigma of Human and Great Ape Evolution," *The Anatomical Record: Part B: The New Anatomist,* July 2006.

Gross, Richard. *Being Human: Psychological and Philosophical Perspectives.* New York: Routledge, 2013.

Grove, Valerie. "The Born Free Foundation: Still Saving the Lions." *Times Online,* March 14, 2009. Accessed 2009. http://women.timesonline.co.uk/tol/life_and_style/women/the_way_we_live/article5901936.ece.

Groves, Colin. "Why Some Monkeys Are Called Apes." Email to Lynette Shanley, forwarded to primfocus email group, October 9, 2007.

Hanover College. "Body of Liberties" Accessed 2009. http://history.hanover.edu/texts/masslib.htm.

Hauser, Marc D. and Susan Carey. "Spontaneous Representations of Small Numbers of Objects by Rhesus Macaques: Examinations of Content and Format." *Cognitive Psychology*, January 21, 2003. Accessed 2013. http://www.wjh.harvard.edu/~lds/pdfs/hauser2003.pdf.

Häyry, Heta and Matti Häyry. "Who's Like Us?" In *The Great Ape Project: Equality Beyond Humanity*, edited by Paola Cavalieri and Peter Singer, 173-182. New York: St. Martin's Press, 1993.

Helping Hands. "Our History." Accessed 2011. http://www.monkeyhelpers.org/ourhistory/.

Henley, Bill. "Primate Protection. *The Post and Courier*, January 8, 2009. Accessed 2013. http://www.postandcourier.com/news/2009/jan/08/primate_pro tection67601/.

International Primate Protection League. "About IPPL." Accessed 2009. http://www.ippl.org/about.php.

International Primate Protection League. "History: Reasons to Support IPPL." Accessed 2013. http://www.ippl.org/gibbon/about-us/history-reasons-to-support-ippl/.

International Primate Protection League. "US 2010 Primate Imports Increase Slightly Over 2009 Figures." Accessed 2011. https://secure.ippl.org/gibbon-wp/u-s-2010-primate-imports-increase-slightly-over-2009-figures/.

Ivory, Alicia S. "Chimpanzee Laws in the United States and Abroad." Accessed 2012. http://www.animallaw.info/articles/dduschimplaws.htm#I.

Jablon, Robert. "Hollywood Chimps Head to Sanctuary." NBCNews.com, December 11, 2006, Accessed 2013. http://www.msnbc.msn.com/id/16156735/.

Jamieson, Dale. "Great Apes and the Human Resistance to Equality." In *The Great Ape Project: Equality Beyond Humanity*, edited by Paola Cavalieri and Peter Singer, 223-228. New York: St. Martin's Press, 1993.

Jean and Charles Shulz Information Center, Sonoma State University. "The Jack London Online Collection." Accessed 2009. http://london.sonoma.edu/.

Jefferson, Thomas, Letter to Henry Gregoire, February 25, 1809.
          Accessed 2013.
          http://teachingamericanhistory.org/library/document/letter-to-henri-
          gregoire/.

Johnson, Ruthanne. "Behind Closed Doors." *Allanimals,* May/June 2011.

Jones Mark. "Has CITES Had Its Day?." Accessed 2010.
          http://news.bbc.co.uk/2/hi/science/nature/8606011.stm.

Kelly, Walt. "Pogo." *New York Star,* April 22, 1971. Accessed 2013.
          http://en.wikipedia.org/wiki/File:Pogo_-
          _Earth_Day_1971_poster.jpg.

Kennedy, John F. "Moon Speech" speech, Rice Stadium. September 12, 1962,
          National Aeronautics and Space Administration.
          http://er.jsc.nasa.gov/seh/ricetalk.htm.

Klein, Hilton J DVM, MS. "Hazards Associated with the Use of Nonhuman
          Primates in Research." In *Well Being of Nonhuman Primates in
          Research,* edited by Joy A. Mench, D.Phil and Lee Krulisch, 51-54.
          Bethesda, MD: Scientists Center for Animal Welfare, 1990. 51-54.
          http://www.scaw.com/nhp.pdf.

Koko.org (The Gorilla Foundation). "Koko's First Interspecies Web Chat:
          Transcript." Accessed 2013.
          http://www.koko.org/world/talk_aol.html.

Lacey Act of 2008, 16 USC 3371-3378. (2008). Accessed 2013.
          http://www.animallaw.info/statutes/stusfd16usca3371.htm.

Leakey: A Century of the Family in East Africa. "Louis Seymour Bazett
          Leakey." Accessed 2013. http://www.leakey.com/louis_leakey.htm.

Lee, Dan P. "Travis The Menace." *New York Magazine,* January 23, 2011.
          Accessed 2012. http://nymag.com/news/features/70830/.

Linzey, Andrew and Adrian Morrison. "Is It Ever Right For Animals To
          Suffer?" *The Washington Post*, December 27, 2009. Accessed 2013.
          http://www.washingtonpost.com/wp-
          dyn/content/article/2009/12/23/AR2009122301317.html.

Madrigal, Alexis C. "The Horrible Thing That Happened to Enos the Chimp When He Orbited Earth 50 Years Ago." *The Atlantic,* November 29, 2011. Accessed 2013. http://www.theatlantic.com/technology/archive/2011/11/the-horrible-thing-that-happened-to-enos-the-chimp-when-he-orbited- earth-50-years-ago/249241/.

Markarian, Michael. "The Great Ape Protection and Cost Savings Act." *Animals & Politics* (blog), *Humane Society Legislative Fund,* April 13, 2011. http://hslf.typepad.com/political_animal/2011/04/the-great-ape-protection-cost-savings-act.html.

Marks, David. "Monkey Helpers Lend a 'Helping Hand'." *8 News Now.* No date. Accessed 2013. http://www.8newsnow.com/story/4361694/monkey-helpers-lend-a-helping-hand?redirected=true.

Massachusetts Society for the Prevention of Cruelty to Animals – Angell Animal Medical Center. "George Thorndike Angell: A Vision Unfolds." Accessed 2013.http://www.mspca.org/about-us/history/george-angell.html.

Maslin, Sarah. "A Tighter Leash on Exotic Pets." *The New York Times,* January 10, 2012. Accessed 2012. http://www.nytimes.com/2012/01/11/us/exotic-animals-business-faces-restrictions.html?_r=2&pagewanted=all%3Fsrc%3Dtp&smid=fb-share.

McClatchy. "Chimps: Life in the Lab." Accessed 2013. http://www.mcclatchydc.com/chimps/.

McGreal, Shirley. "Bangkok Six – 20th Anniversary." Email to primfocus email group, February 20, 2010.

Messenger, Stephen. "Are Zoos Prisons? Habeas Corpus Filed for Chimp."*Treehugger*, January 11, 2010. Accessed 2013. http://www.treehugger.com/corporate-responsibility/are-zoos-prisons-habeas-corpus-filed-for-chimp.html.

Meyer, Tara. "Chantek Signs, Apes Humans." *The Associated Press,* November 27, 1997. http://www.primatesworld.com/SigningOrangutan.html.

Miles, H. Lyn White. "Language and the Orang-utan: In *The Great Ape Project: Equality Beyond Humanity*, edited by Paola Cavalieri and Peter Singer, 45-57. New York: St. Martin's Press, 1993.

Miller, Harlan. "The Wahokies." In *The Great Ape Project: Equality Beyond Humanity*, edited by Paola Cavalieri and Peter Singer, 230-237. New York: St. Martin's Press, 1993.

Mirriam-Webster. "Vivisection." Accessed 2013. http://www.mirriam-webster.com/dictionary/vivisection.

Mitchell, Robert W. "Humans, Nonhumans and Personhood." In *The Great Ape Project: Equality Beyond Humanity*, edited by Paola Cavalieri and Peter Singer, 237-247. New York: St. Martin's Press, 1993.

Nagell, K., R.S. Olguin, and M. Tomasello. "Processes of Social Learning in the Tool Use of Chimpanzees (Pan troglodytes) and Human Children (Homo sapiens)." *Journal of Comparative Psychology,* 1993.

National Association for Biomedical Research. "Overview of Existing System." Accessed 2011. http://www.nabranimallaw.org/Research_Animal_Protection/Overview_of_Existing_System/.

National Association for Biomedical Research. "Species in Research." Accessed 2010. http://www.nabr.org/Biomedical_Research/Laboratory_Animals/Species_in_Research.aspx.

National Geographic Channel. "Chimps On The Edge." National Geographic Channel video. Accessed 2009. http://channel.nationalgeographic.com/channel/videos/chimps-on-the-edge/.

National Institutes of Health. "New Genome Comparison Finds Chimps, Humans Very Similar at the DNA Level." Accessed 2009. http://www.nih.gov/news/pr/aug2005/nhgri-31.htm.

National Research Council (U.S.) *The Psychological Well-Being of Nonhuman Primates.* Washington, D.C.: National Academies Press, 1998. http://www.nap.edu/catalog.php?record_id=4909#toc.

New England Anti-Vivisection Society. "About NEAVS, 1895-1920: The Beginnings – Sowing The Seeds." Accessed 2011. http://www.neavs.org/aboutneavs/history_1895_1920.htm.

New England Primate Sanctuary. "New England Primate Sanctuary:
       FAQs." Accessed 2010.
       http://neprimatesanctuary.org/faqs.html#Howmanyprimatesneed
       sanctuary.

"New EU Rules on Animal Testing Ban Use of Apes." *Independent,*
       September 12, 2010. Accessed 2013.
       http://www.independent.co.uk/life-style/health-and-families/new-eu-
       rules-on-animals-testing-ban-use-of-apes-20077443.html.

Noske, Barbara. "Great Apes as Anthropological Subjects – Deconstructing
       Anthropocentrism." In *The Great Ape Project: Equality Beyond
       Humanity*, edited by Paola Cavalieri and Peter Singer, 258-268. New
       York: St. Martin's Press, 1993.

NOVA. "Ape Genius." NOVA video. Accessed 2008.
       http://pbs.org/wgbh/nova/nature/ape-genius.html.

NOVA. "First Primates: Expert Q&A (Dr. Mary T. Silcox.)" Accessed 2013.
       http://www.pbs.org/wgbh/nova/evolution/first-primates-expert-
       q.html.

O'Carroll, Eoin. "Spain to Grant Some Human Rights to Apes." *The
       Christian Science Monitor,* June 27, 2008.
       http://www.csmonitor.com/Environment/Bright-
       Green/2008/0627/spain-to-grant-some-human-rights-to-apes.

O'Neill, Anne Marie, "One Great Ape". *People,* September 2, 1996.
       Accessed 2013.
       http://www.people.com/people/archive/article/0,,20142136,00.html.

Office of Science and Technology Policy. "U.S. Government Principles
       for the Utilization and Care of Vertebrate Animals Used in
       Testing, Research, and Training." *Federal Register*, 20 May, 1985.
       http://fmp-8.cit.nih.gov/oacu/guidepi/references/2govprinciple.pdf.

One Small Step: The Story of the Space Chimps. Directed by David
       Cassidy and Kristin Davy. Gainesville, FL: The Documentary
       Institute, 2002. DVD.

One Small Step: The Story Of The Space Chimps. "One Small Step: The Story
       of the Space Chimps – Their Story." Accessed 2013.
       http://spacechimps.com/theirstory.html.

Orangutan Foundation International. "Dr. Birute Mary Galdikas Bio." Accessed 2011. http://www.orangutan.org/dr-galdikas-bio.

Orangutan Foundation International. "History of OFI." Accessed 2011. http://www.orangutan.org/about-ofi/history-of-ofi.

Orangutan Foundation International. "Rehabilitation." Accessed 2011. http://www.orangutan.org/our-projects/rehabilitation.

Orangutan Outreach. "About Orangutans." Accessed 2013. http://redapes.org/about-orangutans/.

Orangutan Outreach. "Nyaru Menteng." Accessed 2013. http://redapes.org/bos-projects/nyaru-menteng/.

Organization of American Historians. *Proceedings of the Mississippi Valley Historical Association: Volume X Part II for the Year 1919-1920.* Cedar Rapids: The Torch Press, 1921. Accessed 2013. http://books.google.com/books?id=NxkFAAAAYAAJ&printsec=frontcover#v=oˆnepage&q=&f=false.

Orzech, Kathryn. "What Makes a Primate a Primate?" Accessed 2013. http://tolweb.org/treehouses/?treehouse_id=3029.

Owens, Nick. "Shocking Truth of How Monkeys Are Tortured For 'Entertainment' in Indonesia". *Mirror,* February 6, 2011. Accessed 2013.   http://www.mirror.co.uk/news/top-stories/2011/02/06/shocking-truth-of-how-monkeys-are-tortured-for-entertainment-in-indonesia-115875-22900916/.

Patterson, Francine and Wendy Gordon. "The Case for the Personhood of Gorillas." In *The Great Ape Project: Equality Beyond Humanity*, edited by Paola Cavalieri and Peter Singer, 58-77. New York: St. Martin's Press, 1993.

Peterson, Dale. *Jane Goodall: The Woman Who Redefined Man.* New York: Houghton Mifflin Harcourt, 2006.

Peterson, Dale and Jane Goodall. *Visions of Caliban: On Chimpanzees and People.* Athens: University of Georgia Press, 2000.

Pet Monkey Info. "PetMonkeyInfo.com." Accessed 2011. http://petmonkeyinfo.com/.

Petmonkeyinfo.org. "The Phenomenon of Monkeys as 'Surrogate Children'."
    Accessed 2011.
    http://petmonkeyinfo.org/Monkeys_as_Surrogate_Children.pdf.

Pollard, Katherine S., Sofie R. Salama, Bryan King, Andrew D. Kern, et al.
    "Forces Shaping the Fastest Evolving Regions In The Human
    Genome." PLOS Genetics. Accessed 2013.
    http://www.plosgenetics.org/article/info%3Adoi%2F10.1371%
    2Fjournal.pgen.0020168.

Porton, Ingrid, Scott Carter, Benjamin Beck, and Andy Baker. "Bedtime for
    Bonzo: The Real Bedtime Story." Accessed 2013.
    http://www.honoluluzoo.org/pets.htm.

PressTV. "Female Chimps Play with Stick Dolls." Last modified
    December 26, 2010. http://www.presstv.ir/detail/157239.html.

Primate Rescue Center. "Primate Rescue Center FAQs." Accessed 2009.
    http://www.primaterescue.org/about/faqs.php4.

Primate Rescue Center. "Rescues - The Dahlonega Five." Accessed 2009.
    http://www.primaterescue.org/residents/rescues/dahlonega.php4.

Primate Rescue Center. "The Dahlonega Five." Accessed 2010.
    http://www.primaterescue.org/index.php/our-residents/rescue-
    stories/the-dahlonega-five.

Primates.com. "Order Primates." Accessed 2011.
    http://www.primates.com/primate/index.html.

Project R&R. "Chronology of Key Events in the Scientific Use of
    Chimpanzees in the U.S." Accessed 2011.
    http://www.releasechimps.org/2011/03/02/chronology-of-key-events-
    in-the-scientific-use-of-chimpanzees-in-the-us/#axzz1KxudKkkG.

Project R&R. "End Chimpanzee Research: An Overview." Accessed
    2011.http://www.releasechimps.org/mission/end-chimpanzee-
    research/#axzz1KxudKkkG.

Project R&R. "International Bans." Accessed 2011.
    http://www.releasechimps.org/mission/end-chimpanzee-
    research/country-bans.

Project R&R. "The CHIMP Act." Accessed 2011.
http://www.releasechimps.org/mission/change-laws/the-chimp-act/.

Project R&R. "The Great Ape Protection and Cost Savings Act
(H.R.1313/S.810)." Accessed 2011.
http://www.releasechimps.org/mission/change-laws/the-great-ape-protection-act/#axzz1Lu362lbQ.

Quammen, David. "Jane: Fifty Years at Gombe." *National Geographic,*
October 2010, 110-129.

Rachels, James. "Why Darwinians Should Support Equal Treatment for Other
Great Apes." In *The Great Ape Project: Equality Beyond Humanity*,
edited by Paola Cavalieri and Peter Singer, 152-157. New York: St.
Martin's Press, 1993.

Raffaele, Paul. "Speaking Bonobo." *Smithsonian Magazine,* November 2006.
http://www.smithsonianmag.com/science-nature/10022981.html.

Rees, Amanda. "Reflections on the Field: Primatology, Popular Science, and
the Politics of Personhood." *Social Studies of Science,* December
2007: 881-907.

Regan, Tom. "Ill-gotten Gains." In *The Great Ape Project: Equality Beyond
Humanity*, edited by Paola Cavalieri and Peter Singer, 194-206. New
York: St. Martin's Press, 1993.

Rice, Stanley A. *Encyclopedia of Evolution.* New York: Facts on File
Incorporated, 2011.

Richey, Warren. "Supreme Court Animal Cruelty Ruling: All Sides Find
Positives." *The Christian Science Monitor,* April 20, 2010.
Accessed 2013.
http://www.csmonitor.com/USA/Justice/2010/0420/Supreme-Court-animal-cruelty-ruling-All-sides-find-positives.

Riley, Erin P. "Ethnoprimatology: Toward Reconciliation of Biological and
Cultural Anthropology." *Ecological and Environmental
Anthropology*, 2006. Accessed 2013.
http://eea.uga.edu/02_2006/pdfs/riley_2006.pdf.

Ristau, Carolyn A. and Donald Robbins. "Language in the Great Apes."
Advances in the Study of Behavior, Vol. 12, edited by Jay S.
Rosenblatt. Waltham, Massachusetts: Academic Press, 1982.

Rollin, Bernard. "The Ascent of Apes – Broadening the Moral Community." In *The Great Ape Project: Equality Beyond Humanity*, edited by Paola Cavalieri and Peter Singer, 206-219. New York: St. Martin's Press, 1993.

Rowe, Noel. *The Pictorial Guide to the Living Primates*. Charlestown: Pogonias Press, 1996.

Ryder, Richard D. "Sentientism." In *The Great Ape Project: Equality Beyond Humanity*, edited by Paola Cavalieri and Peter Singer, 220-222. New York: St. Martin's Press, 1993.

Sapontzis, Steve F. "Aping Persons – Pro and Con." In *The Great Ape Project: Equality Beyond Humanity*, edited by Paola Cavalieri and Peter Singer, 269-277. New York: St. Martin's Press, 1993.

SATYA. "NYU Timeline" Accessed 2013. http://web.archive.org/web/20070927024245/http://www.satya mag.com/apr98/nyu_timeline.html.

Savage-Rumbaugh, Susan. "Susan Savage-Rumbaugh on Apes." Filmed February 2004. TED video, 17:27. Posted April 2007. http://www.ted.com/talks/susan_savage_rumbaugh_on_apes_th at_write.html.

Save The Chimps. "Chimps in Space." Accessed 2013. http://www.savethechimps.org/chimps-in-space.

Save The Chimps. "Rescuing the Coulston Chimps and Transforming a Lab into a Sanctuary." Accessed 2013. http://www.savethechimps.org/rescuing-the-coulston-chimps.

Save The Chimps. "Save The Chimps' Mission." Accessed 2010. http://www.savethechimps.org/about-us.aspx#begining.

Sayers, Ken. "The Chimpanzee Has No Clothes: A Critical Examination of Pan troglodytes in Models of Human Evolution." *Current Anthropology,* February 2008, 87-108.

Seyfarth, Robert M., Dorothy L. Cheney and Peter Marler. "Vervet Monkey Alarm Calls: Semantic Communication in a Free-Ranging Primate." *Animal Behavior,* November 1980, 1070-1094. Accessed 2009. http://www.sciencedirect.com/science/article/pii/S0003347280 800972.

Science Daily. "Comparing Chimp, Human DNA." Accessed 2013.
    http://www.sciencedaily.com/releases/2006/10/061013104633.
    htm.

Science Daily. "Fossil Discovery: More Evidence for Asia, Not Africa, as the
    Source of Earliest Anthropoid Primates." Accessed 2013.
    http://www.sciencedaily.com/releases/2012/06/120604155705.
    htm.

Science Daily. "Fossil Teeth Reveal Oldest Bushbabies, Lorises."
    Accessed 2013.
    http://www.sciencedaily.com/releases/2003/04/030402073345.
    htm.

Science Daily. "Recent Census in War-Torn DR Congo Finds Gorillas Have
    Survived, Even Increased." Accessed 2013.
    http://www.sciencedaily.com/releases/2011/04/110414141406.
    htm.

Scientific American. "Ban Chimp Testing." September 28, 2011.
    http://www.scientificamerican.com/article.cfm?id=ban-chimp-
    testing&WT.mc_id=SA_emailfriend.

Siebert, Charles. *The Wauchula Woods Accord: Toward a New Understanding
    of Animals.* New York: Scribner, 2009.

Sinclair, Upton. *The Jungle.* New York: Bantam Books, 1981.

Singer, Peter. *Animal Liberation.* New York: HarperCollins Publishers, 2002.

Skloot, Rebecca. "Creature Comforts". *The New York Times Magazine,*
    December 31,2008. Accessed 2013.
    http://www.nytimes.com/2009/01/04/magazine/04Creatures-
    t.html?_r=1.

Smith, Rob Roy. "Standing on Their Own Four Legs: The Future of Animal
    Welfare Litigation After Animal Legal Defense Fund, Inc. v.
    Glickman". *Environmental Law*, Winter 1999.
    http://nationalaglawcenter.org/assets/bibarticles/smith_four.pdf.

Smithsonian National Museum of Natural History. "Australopithecus
    africanus." Accessed 2013.
    http://humanorigins.si.edu/evidence/humanfossils/species/austr
    alopithecus-africanus.

"Spanish Parliament to Extend Rights to Apes." *Reuters,* June 25, 2008. Accessed 2011. http://www.reuters.com/article/2008/06/25/us-spain-apes-idUSL256586320080625.

Spokane Daily Chronicle. "Problems Told: 'Bugs' Plagued Space Trip." *Spokane Daily Chronicle,* February 2, 1961. Accessed 2013. http://news.google.com/newspapers?nid=1338&dat=19610202 &id=WpUSAAAAIBAJ&sjid=PPcDAAAAIBAJ&pg=3862,68 1440.

Stanford, Craig. *Significant Others: the Ape-Human Continuum and the Quest for Human Nature.* New York: Basic Books, 2001.

Stauffer, R.L., A. Walker, et al. "Human and Ape Molecular Clocks and Constraints on Paleontological Hypotheses." *Journal of Heredity,* November 2001. http://jhered.oxfordjournals.org/content/92/6/469.long.

Steinberg, Brian. "How 'Chimpus Commercialus' Went From Ad Star to Endangered Species". *AdvertisingAge,* February 6, 2011. Accessed 2011. http://adage.com/article/special-report-super-bowl/chimps-apes-ad-star-endangered-species/148714/.

Stop Animal Exploitation NOW! "Stop Animal Exploitation NOW!" Accessed 2010. http://www.all-creatures.org/saen/.

Student Animal Legal Defense Fund. "Law Students & SALDF Chapters." Accessed 2009. http://www.saldf.org.

Sumatran Orangutan Society. "Orangutan Facts." Accessed 2010. http://www.orangutans-sos.org/kids/orangutan_facts/.

Sussman, R.W., Paul A. Garber, and Jim M. Cheverud. "Importance of Cooperation and Affiliation in the Evolution of Primate Sociality." *American Journal of Physical Anthropology,* September 2005. Accessed 2013. http://www.ncbi.nlm.nih.gov/pubmed/15778981.

Tanner, Leslie, ed., *Voices From Women's Liberation.* New York: Signet, 1970.

Teleki, Geza. "They Are Us." In *The Great Ape Project: Equality Beyond Humanity,* edited by Paola Cavalieri and Peter Singer, 296-302. New York: St. Martin's Press, 1993.

Temerlin, Maurice K. *Lucy: Growing Up Human*. Palo Alto, California: Science and Behavior Books, Inc, 1975.

Terrace, Herbert S. *Nim: A Chimpanzee who Learned Sign Language*. New York: Washington Square Press, 1979.

The British Union for the Abolition of Vivisection. "BUAV Primate Trade Investigations." Accessed 2011. http://www.buav.org/our-campaigns/primate-campaign/buav-primate-trade-investigations/.

The British Union for the Abolition of Vivisection. "Our Achievements." Accessed 2011. http://www.buav.org/about-us/our-achievements/.

The British Union for the Abolition of Vivisection. "Our History." Accessed 2011. http://www.buav.org/about-us/our-history/.

The British Union for the Abolishment of Vivisection. "PACE (People Against Chimpanzee Experiments) Merges with the BUAV." Accessed 2013. http://www.buav.org/article/644/pace-people-against-chimpanzee-experiments-merges-with-the-buav.

The Dian Fossey Gorilla Fund International. "Dian Fossey – Biography." Accessed 2013.http://gorillafund.org/page.aspx?pid=380.

The Dian Fossey Gorilla Fund International. "Karisoke Research Center." Accessed 2011. http://gorillafund.org/Page.aspx?pid=381.

The Dian Fossey Gorilla Fund International. "The Fossey Fund's Work In Congo." Accessed 2011. http://gorillafund.org/Page.aspx?pid=267.

The Dian Fossey Gorilla Fund International. "The GRACE Center for Rescued Gorillas." Accessed 2011. http://gorillafund.org/Page.aspx?pid=254.

The Gorilla Foundation. "Gorilla Foundation – Preservation of Gorillas." Accessed 2010.   http://www.koko.org/foundation/.

The Humane Society of the United States. "The HSUS Applauds Signing of Animal Crush Video Prohibition Act." Accessed 2010. http://www.humanesociety.org/news/press_releases/2010/12/crush_bill_signed_120910.html.

The IUCN Red List of Endangered Species. "Pan Paniscus." Accessed 2013. http://www.iucnredlist.org/details/15932/0.

The Jane Goodall Foundation. "About JGI." Accessed 2009.
    http://www.janegoodall.org/about-jgi.

The Jane Goodall Institute. "Early Days." Accessed 2009.
    http://www.janegoodall.org/janes-story.

The Jane Goodall Institute. "Gombe Q&A." Accessed 2013.
    http://www.janegoodall.org/gombe50/faq2.

The Jane Goodall Institute. "Rescuing Orphaned Chimpanzees."
    Accessed 2009. http://www.janegoodall.org/chimpanzees-rescue.

The National Academies Press. "Chimpanzees in Biomedical and Behavioral
Research: Assessing the Necessity." Accessed 2011.
    http://books.nap.edu/openbook.php?record_id=13257&page=R1.

The New York Times. "Passion For Animals Really a Disease." *The New York
    Times,* March 8, 1909. Accessed 2013.
    http://query.nytimes.com/mem/archive-
    free/pdf?res=FA0910F73B5512738DDDA10894DB405B898C
    F1D3.

The Oklahoma State Courts Network. "Depiction of Animal Cruelty."
    Accessed 2013.
    http://www.oscn.net/applications/oscn/DeliverDocument.asp?C
    iteID=198777.

The Post and Courier. "Dr. McGreal's Deserved Honor." *The Post and
    Courier*, January 4, 2008. Accessed 2011.
    http://www.postandcourier.com/news/2008/jan/04/dr_mcgreals
    _deserved_honor26596/.

The Ram Dass Library. "Hanuman." Accessed 2013.
    http://www.ramdasstapes.org/hanuman.htm.

Thirlway, Helen. "CITES (and IPPL) at 40: Insights from Bangkok." *IPPL
    News,* April 2013.

Thomas, Heather. "Lemurs as Pets." Accessed 2009.
    http://petmonkeyinfo.org/prosimians.htm.

Thompson, Shawn. *The Intimate Ape.* New York: Citadel Press, 2010.

Thompson, Shawn. "Why Chimpanzees Would Dance to Johnny Cash's Music." *Psychology Today,* November 13, 2010. Accessed 2011. http://www.psychologytoday.com/blog/the-intimate-ape/201011/why-chimpanzees-would-dance-johnny-cashs-music.

Thoreau, Henry David. *Walden.* New York: Thomas Y. Crowell and Company, 1910. Accessed 2011. http://books.google.com/books?id=yiQ3AAAAIAAJ&printsec =titlepage&source=gbs_navlinks_s#v=onepage&q=&f=false.

Trottier, Alexandra. "Animal Cruelty in Hollywood: The Chimpanzee" *Culture Magazine,* August 29, 2009. Accessed 2010. http://culturemagazine.ca/cinema/animal_cruelty_in_hol d_the_chimpanzee.html.

Tufts University. "Daniel C. Dennett's Home Page." Accessed 2010. http://ase.tufts.edu/cogstud/incbios/dennettd/dennettd.htm.

Tufts University School of Veterinary Medicine, Center for Animals and Public Policy. "The Animal Policy Report." December 1998. http://petmonkeyinfo.org/position.htm.

U.S. Department of Health & Human Services, National Institutes of Health, Office of Extramural Research. "Health Research Extension Act of 1985." Accessed 2013. http://grants.nih.gov/grants/olaw/references/hrea1985.htm.

U.S. Fish and Wildlife Service Virtual Newsroom. "U.S. Fish and Wildlife Service Initiates Review of the Chimpanzee's Status." Accessed 2011. http://us.vocuspr.com/Newsroom/Query.aspx?SiteName=fws &Entity=PRAsset&SF_PRAsset_PRAssetID_EQ=128219& XSL=PressRelease&Cache=True.

U.S. Food and Drug Administration. "Animal Testing." Accessed 2012. http://www.fda.gov/Cosmetics/ProductandIngredientSafety/Pro ductTesting/ucm072268.htm.

"U.S. Government Principles for the Utilization and Care of   Vertebrate Animals Used in Testing, Research and Training." *Federal Register,* May 20, 1985. http://fmp8.cit.nih.gov/oacu/guidepi/references/2govprinciple.pdf.

U.S. Senate Committee on Environment and Public Works. "Endangered Species Act of 1973." Accessed 2013. http://www.epw.senate.gov/esa73.pdf.

United Nations. "World Charter for Nature." October 28, 1982.
    Accessed 2013.
    http://www.un.org/documents/ga/res/37/a37r007.htm.

United Nations Environment Programme. "Meetings: Intergovernmental
    Meeting on Great Apes." Accessed 2010.
    http://www.unep.org/grasp/Meetings/IGM-kinshasa/Outcomes/index-
    reports.asp.

United States Department of Agriculture, Animal and Plant Inspection
    Service. "Animal Welfare." Accessed 2009.
    http://www.aphis.usda.gov/animal_welfare/index.shtml.

United States Department of Agriculture, Animal and Plant Health Inspection
    Service. "Annual Report Animal Usage by Fiscal Year, 2009".
    Accessed 2013.
    http://www.aphis.usda.gov/animal_welfare/efoia/downloads/20
    09_Animals_Used_In_Research.pdf.

United States Department of Agriculture, Animal and Plant Health Inspection
    Service. "Annual Report Animal Usage by Fiscal Year, 2010."
    Accessed 2013.
    http://www.aphis.usda.gov/animal_welfare/efoia/downloads/20
    10_Animals_Used_In_Research.pdf.

United States Department of Agriculture, National Agriculture Library.
    "Environmental Enrichment for Nonhuman Primates Resource
    Guide." Accessed 2011.
    http://www.nal.usda.gov/awic/pubs/Primates2009/primates.shtml.

USLegal. "Habeas Corpus Law & Legal Definition." Accessed 2010.
    http://definitions.uslegal.com/h/habeas-corpus.

Van Schaik, Carel P. et al. "Orangutan Cultures and the Evolution of
    Material Culture." *Science,* January 2003. Accessed 2012.
    http://www.sciencemag.org/content/299/5603/102.full.pdf.

Wade, Nicholas. "Deciphering the Chatter of Monkeys and Chimps." *The New
    York Times,* January 11, 2010. Accessed 2013.
    http://www.nytimes.com/2010/01/12/science/12monkey.html?_
    r=1&adxnnl=1&adxnnlx=1292695596-
    7fryqkbjBIgO0drYJMeTBA.

Walsh, Bryan. "Why The Stamford Chimp Attacked." *Time Magazine,* February 18, 2009. Accessed 2011. http://www.time.com/time/health/article/0,8599,1880229,0 0.html.

Warner, Margaret. "Supreme Court Overturns Law Banning Videos Depicting Animal Cruelty." *PBS NewsHour,* April 20, 2010. Accessed 2013. http://www.pbs.org/newshour/bb/law/jan-june10/scotus_04-20.html.

Widowski, Tina. "The Evaluation and Promotion of Well-being in Farm Animals and Laboratory Primates: Common Problems in *Primates in Research*, edited by Joy A. Mench, D. Phil and Lee Krulisch, 23-28. Bethesda, MD: Scientists Center for Animal Welfare, 1990. http://www.scaw.com/nhp.pdf.

Wilcox, Ella Wheeler. *The Voice of the Voiceless, Ver.II.* Animal Liberation Front. Accessed 2009. http://www.animalliberationfront.com/Saints/Authors/Poetry/V oice_of_Voiceless.htm.

Wildlife Conservation Society. "Mountain Gorilla." Accessed 2013. http://www.wcs.org/saving-wildlife/great-apes/mountain-gorilla.aspx.

Wise, Steven M. *Rattling The Cage: Toward Legal Rights for Animals.* New York: Perseus Publishing, 2000.

World Association of Zoos and Aquariums. "Zoos and Aquariums of the World." Accessed 2010. http://www.waza.org/en/site/zoos-aquariums.

Wynne, Clive. "Aping Language: A Skeptical Analysis of the Evidence for Nonhuman Primate Language." *Skeptic,* October 31, 2007. Accessed 2011. http://www.skeptic.com/eskeptic/07-10-31/#feature.

Ynterian, Pedro A. "Why Zoonit Would Be Different*?"* *Great Ape Protection,* October 26, 2010. Accessed 2011. http://www.greatapeproject.org/en-US/noticias/Show/3340,why-zoonit-would-be-different.

Yule, George. *The Study of Language.* New York: Cambrige University Press, 2010.

Zig Zag Productions. "My Child Is A Monkey Part 1." YouTube video, 2009.
        http://www.youtube.com/watch?v=vx1OQ_x9tFk&feature=mf
        u_in_order&list=UL.

Zig Zag Productions. "My Child Is A Monkey Part 2." YouTube video, 2009.
        http://www.youtube.com/watch?v=DbrIuZ2YFic.

Zig Zag Productions. "My Child Is A Monkey Part 3." YouTube video, 2009.
        http://www.youtube.com/watch?v=nxkvvnwcaNk&NR=1.

Zig Zag Productions. "My Child Is A Monkey Part 4." YouTube video, 2009.
        http://www.youtube.com/watch?v=sCkNkGJDb6A&NR=1&fe
        ature=fvwp.

Zig Zag Productions. "My Child Is A Monkey Part 5." YouTube video, 2009.
        http://www.youtube.com/watch?v=AFr3Kr6ov88.

Zig Zag Productions. "My Child Is A Monkey Part 6." YouTube video, 2009.
        http://www.youtube.com/watch?v=wrTpUzIZ_zs&NR=1&feat
        ure=fvwp.

Index

Notes

[1] De Waal and Lanting, *Bonobo*, 6

[2] Pollard et al, "Forces."

[3] Wise, *Rattling*, 2000, 132.

[4] Favre and Tsang. "Development."

[5] Francione, *Animals as Persons*, xi.

[6] Francione, *Animals, Property*, 34.

[7] Hanover, "Massachusetts."

[8] Organizaton of American Historians, *Proceedings*, 281.

[9] Favre and Tsang. "Development."

[10] Beers, *Prevention*, 311.

[11] American Rights History, "Cruel."

[12] Favre and Tsang. "Development."

[13] Beers, *Prevention*, 32.

[14] Ibid.

[15] Ibid., 24.

[16] Ibid.

[17] Sinclair, *Jungle*, 33.

[18] Thorcau, *Walden*, 3.

[19] Beers, *Prevention*, 3.

[20] Animal Rights History, *Cruelty*.

[21] Beers, *Prevention,* 44.

[22] Ibid., 15.

[23] Ibid., 61.

[24] The New York Times, "Passion."

[25] Beers, *Prevention,* 12.

[26] Ibid., 67.

[27] American Humane Association, *Who.*

[28] Beers, *Prevention,* 70.

[29] Ibid., 73.

[30] Bideawee, *History.*

[31] Beers, *Prevention,* 75-79.

[32] Ibid., 78.

[33] Ibid., 81.

[34] Beers, *Prevention,* 83.

[35] Beers, *Prevention,* 86.

[36] Massachusetts Society, *George.*

[37] Beers, *Prevention,* 88-89.

[38] Ibid., 99-100.

[39] Ibid., 100.

[40] Ibid., 101.

[41] Ibid., 105-107.

[42] Ibid., 104.

[43] Ibid., 109.

[44] Ibid., 113.

[45] Ibid., 115.

[46] Mirriam-Webster, *Vivisection.*

[47] Beers, *Prevention,* 122.

[48] Ibid., 124.

[49] Ibid., 126.

[50] Blum, *Monkey,* 109.

[51] Beers, *Prevention,* 139.

[52] Ibid., 135.

[53] Ibid., 136.

[54] Ibid., 155.

[55] Ibid., 163.

[56] Ibid., 164-165.

[57] Ibid., 177.

[58] Blum, *Monkey,* 43.

[59] Favre, "Overview."

[60] Beers, *Prevention,* 174.

[61] Ibid., 193.

[62] U.S. Senate, "Endangered."

[63] Convention, "Convention."

[64] Beers, *Prevention,* 178-179.

[65] Ibid.

[66] Blum, *Monkey,* 113.

[67] Teleki, "They," 298.

[6868] Orzech, "What."

[69] National Research, *Psychological,* 7.

[70] Ibid., 42.

[71] Ibid., 7-8.

[72] Ibid., 82.

[73] Ibid., 92.

[74] Ibid., 8.

[75] Ibid., 7.

[76] De Waal, *Family,* 18.

[77] Cohen, "Thinking," 50-57.

[78] De Waal, *Family,* 27, 94.

[79] Ibid., 25.

[80] Cohen, "Thinking," 50-57.

[81] De Waal, *Family,* 40.

[82] BBC News, "Chimps."

[83] Cavalieri and Singer, *Great,* 96.

[84] De Waal and Lanting, *Bonobo*, 26.

[85] Ibid.

[86] Sussman, Garber, Cheverud, "Importance."

[87] De Waal and Lanting, *Bonobo*, 30.

[88] Gardner, "Smiling."

[89] National Research, *Psychological,* 77.

[90] De Waal, *Family,* 136.

[91] Peterson and Goodall, *Visions,* 21.

[92] De Waal, "Empathy," .87-106.

[93] De Waal, *Primates,* 30.

[94] O'Neill, "One."

[95] Thompson, *Intimate,* 116.

[96] Ibid., 170.

[97] Ibid., 126.

[98] Ibid., 117.

[99] Ibid., 119.

[100] Ibid., 171.

[101] Boyd Group, "Paper 2."

[102] Wise, *Rattling,* 134.

[103] Ibid.

[104] Nagell, Olguin, and Tomasello. "Processes," 174-186.

[105] Thompson, *Intimate,* 125.

[106] Wise, *Rattling,* 183.

[107] Ibid., 184.

[108] Ibid., 185.

[109] Ibid.,186.

[110] Ibid., 187.

[111] Grehan, "Mona."

[112] Cohen, "Thinking," 50-57.

[113] Ibid.

[114] Wise, *Rattling,* 190.

[115] De Waal, *Primates,* 43-44.

[116] Hauser and Carey. "Spontaneous."

[117] van Schaik et al., "Orangutan," 102.

[118] Ibid., 105.

[119] Sayers, "Chimpanzee," 87-108.

[120] Noske, "Great," 259.

[121] van Schaik et al., "Orangutan," 103.

[122] Wise, *Rattling,* 180.

[123] Sayers, "Chimpanzee," 87-108.

[124] De Waal, *Family,* 145.

[125] NOVA, "Ape," 11:05.

[126] Wise, *Rattling,* 192.

[127] NOVA, "Ape," 5:13.

[128] Diamond, "Third," 71.

[129] Boyd Group, "Paper 2."

[130] National Research, *Psychological,* 77.

[131] Blum, *Monkey,* 7.

[132] Peterson and Goodall, *Visions,* 23.

[133] Ibid., 164.

[134] PressTV, "Female."

[135] Thompson, "Chimpanzees."

[136] Stanford, *Significant.*

[137] De Waal and Lanting, *Bonobo,* 118.

[138] Rice, *Encyclopedia,* 107.

[139] Smithsonian, "*Australopithecus.*"

[140] De Waal and Lanting, *Bonobo,* 1-3.

[141] Ibid., 112.

[142] Ibid., 4.

[143] Ibid., 65.

[144] Ibid., 66.

[145] Ibid., 84-85.

[146] The IUCN, "Pan Paniscus."

[147] Cohen, "Thinking," 50-57.

[148] Ibid.

[149] Noske, "Great," 265-266.

[150] Ibid., 259.

[151] NOVA, "Ape," 21:47.

[152] Terrace, *Nim,* 13.

[153] Miles, "Language," 46.

[154] Gross, *Being*. Page unknown.

[155] Temerlin, *Lucy*.

[156] Peterson and Goodall, *Visions,* 207.

[157] Ibid., 208.

[158] Wade, "Deciphering."

[159] Peterson and Goodall, *Visions,* 213-214.

[160] Ibid., 215.

[161] Terrace, *Nim,* 15.

[162] Ristau and Robbins. "Language," 163.

[163] Fouts and Fouts, "Chimpanzees'," 29.

[164] Friends of Washoe, "Tributes." 2013.

[165] Fouts and Mills, *Next,* 310-343.

[166] Ibid., 242.

[167] Ibid., 300.

[168] Ibid., 301.

[169] Fouts and Fouts, "Chimpanzees'," 35.

[170] Blum, *Monkey,* 15.

[171] Terrace, *Nim,* 23-35.

[172] Ibid., 137.

[173] Ibid., 250.

[174] Ibid., 111-112.

[175] Ibid., 116.

[176] Ibid., 120.

[177] Ibid., 143.

[178] Ibid., 129-130.

[179] Ibid., 210.

[180] Ibid., 210-215.

[181] Wise, *Rattling,*173.

[182] Peterson and Goodall, *Visions,* 220-221.

[183] Koko.org, "Koko's."

[184] Patterson and Gordon. "Case," 61.

[185] Ibid., 59.

[186] Ibid., 67.

[187] Ibid., 59-60.

[188] Ibid., 62.

[189] Ibid.

[190] Ibid.

[191] Ibid., 64.

[192] Ibid.

[193] Ibid., 62.

[194] Ibid., 64.

[195] Ibid., 65.

[196] Ibid., 62.

[197] Ibid., 66.

[198] Anderson, *Doctor,* 284.

[199] Miles, "Language," 42.

[200] Ibid., 47.

[201] Ibid., 48.

[202] Ibid., 48-50.

[203] Ibid., 48.

[204] Ibid., 49.

[205] Ibid., 50.

[206] Chantek, "Project."

[207] Miles, "Language," 46.

[208] Ibid., 52-54.

[209] Thompson, *Intimate,* 136-146.

[210] Terrace, *Nim,* 19-22.

[211] Raffaele, "Speaking."

[212] NOVA, "Ape," 31:23.

[213] Savage-Rumbaugh, "Susan," 12:04.

[214] Ibid., 13:41.

[215] Wise, *Rattling,* 168.

[216] Terrace, *Nim,* 150.

[217] Fouts and Fouts, "Chimpanzees'," 36.

[218] Terrace, *Nim,* 231-232.

[219] Ibid., 268.

[220] Ibid., 13.

[221] Wise, *Rattling,* 225.

[222] Ibid., 229.

[223] Wynne, "Aping."

[224] Wise, *Rattling,* 226.

[225] Ibid.

[226] Blum, *Monkey,* 19.

[227] Peterson, *Jane,* 612.

[228] Peterson and Goodall, *Visions*, 227.

[229] Wynne, "Aping."

[230] Teleki, "They," 298.

[231] Wise, *Rattling,* 228.

[232] Yule, *Study* 8-12.

[233] De Waal and Lanting, *Bonobo,* 43-44.

[234] Thompson, *Intimate,* 149.

[235] Bekoff, "Common," 105.

[236] Wise, *Rattling,* 220-221.

[237] Cohen, "Thinking," 50-57.

[238] Stanford, *Significant,* 157.

[239] De Waal and Lanting, *Bonobo,* 33.

[240] Wade, "Deciphering."

[241] Seyfarth, Cheney and Marler, "Vervet."

[242] Wade, "Deciphering."

[243] Ibid.

[244] Ibid.

[245] Peterson and Goodall, *Visions*, 180.

[246] BornFree USA, "Summary."

[247] Maslin, "Tighter."

[248] Zig Zag Productions, "Child, Part 4," 0:50 – 1:00.

[249] Blair, "Pets."

[250] Porton, Carter, Beck and Baker. "Bedtime."

[251] Blair, "Pets."

[252] Thompson, *Intimate,* 121.

[253] Peterson and Goodall, *Visions,* 104.

[254] Green and The Center for Public Integrity, *Animal,* 233.

[255] Petmonkeyinfo.org, "Phenomenon.'"

[256] Peterson and Goodall, *Visions,* 154.

[257] Zig Zag Productions, "Child, Part 2," 6:26.

[258] "Chimps" *National Geographic Channel*, 6:51.

[259] Zig Zag Productions, "Child, Part 4," 4:14.

[260] Green and The Center for Public Integrity, *Animal,* 186.

[261] Thomas, "Lemurs."

[262] National Research Council, *Psychological,* 20.

[263] Ibid., 40.

[264] Ibid., 43.

[265] Peterson and Goodall, *Visions,* 143.

[266] Blair, "Pets."

[267] "Chimps" *National Geographic Channel*, 31:20.

[268] Terrace, *Nim,* 130.

[269] Ibid., 144.

[270] Ibid., 144-145.

[271] Thompson, *Intimate,* 219-221.

[272] Lee, "Travis."

[273] Ibid.

[274] "Chimps" *National Geographic Channel*, 32:33.

[275] Porton, Carter, Beck and Baker. "Bedtime."

[276] Green and The Center for Public Integrity, *Animal,* 183.

[277] Ibid.

[278] Ibid.

[279] Ibid., 188.

[280] Tufts, "Animal."

[281] Ibid.

[282] Zig Zag Productions, "Child, Part 2," 0:25-0:32.

[283] Peterson and Goodall, *Visions,* 192.

[284] Primate Rescue Center. "Dahlonega."

[285] Ibid.

[286] Thompson, *Intimate,* 129.

[287] Zig Zag Productions, "Child, Part 5," 1:15.

[288] Ibid., 4:30.

[289] Peterson and Goodall, *Visions,* 81.

[290] Skloot, "Creature."

[291] Ibid.

[292] Helping Hands. "History."

[293] Marks, "Monkey."

[294] Ibid.

[295] De Waal, *Family,* 12-14.

[296] Peterson and Goodall, *Visions,* 180.

[297] De Waal, *Family,* 12-14.16-22.

[298] Peterson and Goodall, *Visions,* 150.

[299] Ibid., 146-150.

[300] Ibid., 148.

[301] Ibid.

[302] Ibid., 149.

[303] Ibid.

[304] Ibid., 174.

[305] Ibid., 157-179.

[306] Ibid., 180.

[307] Baeckler, "Campaign."

[308] Ibid.

[309] Trottier, "Animal."

[310] Baeckler, "Campaign."

[311] Trottier, "Animal."

[312] Ibid.

[313] Steinberg, " 'Chimpus'."

[314] Center for Great Apes, "Feed."

[315] Steinberg, " 'Chimpus'."

[316] Owens, "Shocking."

[317] Center for Great Apes, "Feed."

[318] Steinberg, " 'Chimpus'."

[319] Chimpanzee Sanctuary Northwest, "Missy."

[320] Peterson and Goodall, *Visions,* 82.

[321] National Research Council, *Psychological,* 122.

[322] U.S. Food and Drug Administration, "Animal."

[323] National Association for Biomedical Research, "Species."

[324] Klein, "Hazards," 52.

[325] McClatchy, "Chimps," 2:45.

[326] Singer, *Animal,* 76.

[327] Project R&R, "Chronology."

[328] Blum, *Monkey,* 191.

[329] National Research Council, *Psychological,* 17.

[330] Adams, "Some."

[331] Stop Animal Exploitation NOW!, "Stop."

[332] Finkelmeyer, "Animal."

[333] Wise, *Rattling,* 251-252.

[334] Francione, *Animals, Property,* 167.

[335] Channel3000, "Workers."

[336] Teleki, "They," 300.

[337] Francione, *Animals, Property,* 179-182.

[338] Ibid.

[339] Ibid.

[340] Ibid., 182.

[341] Singer, *Animal,* 27.

[342] *One Small Step: The Story of the Space Chimps,* 12:56.

[343] One Small Step: The Story Of The Space Chimps, "One."

[344] Ibid.

[345] *One Small Step: The Story of the Space Chimps,* 8:50.

[346] One Small Step: The Story Of The Space Chimps, "One."

[347] Ibid.

[348] Save The Chimps, "Chimps."

[349] Spokane Daily Chronicle, "Problems."

[350] *One Small Step: The Story of the Space Chimps,* 27:24.

[351] *One Small Step: The Story of the Space Chimps,* 28:35.

[352] *One Small Step: The Story of the Space Chimps,* 33:20.

[353] Madrigal, "Horrible."

[354] One Small Step: The Story Of The Space Chimps, "One."

[355] One Small Step: The Story Of The Space Chimps, "One."

[356] Kennedy, "Moon."

[357] Ibid.

[358] Ibid.

[359] Save The Chimps, "Chimps."

[360] Correll, "Astrochimps."

[361] One Small Step: The Story Of The Space Chimps, "One."

[362] *One Small Step: The Story of the Space Chimps,* 43:40.

[363] Ibid.

[364] One Small Step: The Story Of The Space Chimps, "One."

[365] Blum, *Monkey,* 42.

[366] Ibid., 39.

[367] Ibid., 107-109.

[368] Ibid., 122.

[369] Ibid.

[370] Ibid., 107-109.

[371] Ibid., 23-24.

[372] Ibid.

[373] Ibid., 82.

[374] Ibid., 83.

[375] Ibid.

[376] Singer, *Animal,* 58.

[377] The National Academies Press, "Chimpanzees," 10.

[378] International Primate Protection League, "2010."

[379] Sayers, "Chimpanzee."

[380] Singer, *Animal,* 89.

[381] Quammen, "Jane."

[382] Blum, *Monkey,* 232.

[383] Altmann, "Introductory," 21-22.

[384] Ibid., 21.

[385] McClatchyy, "Medical," 2:24.

[386] Bekoff, *Animals,* 3.

[387] National Research Council, *Psychological,* 122.

[388] Goodman, "Argument."

[389] Goodman, "Animal.".

[390] The National Academies Press, "Chimpanzees," 2.

[391] Ibid., 4.

[392] Gorman, "U.S."

[393] Gorman, "Agency."

[394] Singer, *Great,* 6.

[395] Blum, *Monkey,* 41.

[396] United States Department of Agriculture, Animal and Plant Health Inspection Service. "Annual...2010."

[397] Kelly, "Pogo."

[398] Brubaker, "Mission."

[399] Ibid.

[400] Ibid.

[401] Thompson, *Intimate,* 35.

[402] Ibid., 25.

[403] Brubaker, "Mission."

[404] Brubaker, "Mission."

[405] Wildlife Conservation Societ, "Mountain."

[406] Science Daily, "Recent."

[407] Stanford, *Significant,* 191-192.

[408] Thompson, *Intimate,* 45.

[409] Blum, *Monkey,* 217.

[410] De Waal and Lanting, *Bonobo,* 173.

[411] Fossey, *Gorillas,* 40.

[412] Peterson and Goodall, *Visions,* 102.

[413] Blum, *Monkey,* 249.

[414] International Primate Protection League, "US."

[415] Thompson, *Intimate,* 191.

[416] Blum, *Monkey,* 249.

[417] Blum, *Monkey,* 250.

[418] Thirlway, "CITES."

[419] Peterson and Goodall, *Visions,* 90.

[420] Thirlway, "CITES."

[421] Jones, "CITES."

[422] Fossey, *Gorillas,* 222.

[423] Ibid., 239.

[424] Ibid., 106-108.

[425] The Dian Fossey Gorilla Fund International, "Dian."

[426] Science Daily, "Recent."

[427] Fossey, *Gorillas,* 242.

[428] Ibid., 57-59.

[429] Stanford, *Significant,* 195.

[430] Peterson and Goodall, *Visions,* 110.

[431] Blum, *Monkey,* 249.

[432] Teleki, "They," 302.

[433] Butler, "Why."

[434] Wise, *Rattling,* 47.

[435] Francione, *Animals, Property,* 121.

[436] Ibid., 125.

[437] Ibid., 95.

[438] Wise, *Rattling,* 25.

[439] Francione, *Animals as Persons,* 61.

[440] Francione, *Animals, Property,* 25.

[441] Ibid., 25.

[442] Wise, *Rattling,* 74.

[443] Francione, *Animals as Persons,* 39-41.

[444] United Nations, "World."

[445] National Association for Biomedical Research, "Overview."

[446] New England Primate Sanctuary. "New."

[447] National Association for Biomedical Research, "Overview."

[448] Lacey Act of 2008, 16 USC 3371- 3378.

[449] Francione, *Animals, Property,* 201.

[450] Favre, "Overview."

[451] Ivory, "Chimpanzee."

[452] Francione, *Animals, Property,* 201.

[453] Widowski, "Evaluation," 23.

[454] Johnson, "Behind."

[455] Francione, *Animals, Property,* 173.

[456] Francione, *Animals, Property,* 205.

[457] Favre, "Overview."

[458] Blum, *Monkey,* 184.

[459] Francione, *Animals, Property,* 186.

[460] Johnson, "Behind."

[461] Ibid.

[462] Ibid.

[463] Ibid.

[464] Smith, "Standing."

[465] U.S. Senate Committee on Environment and Public Works, "Endangered."

[466] Green, and The Center for Public Integrity, *Animal,* 120.

[467] U.S. Fish and Wildlife Service Virtual Newsroom, "U.S. Fish."

[468] Jones, "CITES."

[469] Peterson and Goodall, *Visions,* 104.

[470] CITES, "Mammals."

[471] Peterson and Goodall, *Visions,* 104.

[472] Jones, "CITES."

[473] National Geographic Channel, "Chimps On The Edge," 10:38.

[474] U.S. Department of Health & Human Services, National Institutes of Health, Office of Extramural Research, "Health."

[475] The Humane Society of the United States, "Applauds."

[476] Govtrack.us, "H.R. 5566."

[477] Project R&R, "CHIMP."

[478] Ibid.

[479] Ibid.

[480] Animal Legal & Historical Center, "Chimpanzee."

[481] Ibid.

[482] Chimp Haven, "History."

[483] Animal Legal & Historical Center, "Chimpanzee."

[484] Francione, *Animals as Persons,* 80.

[485] Project R&R, "CHIMP."

[486] Singer, *Animal,* xiii.

[487] Bekoff, *Animals,* 32-33.

[488] Project R&R, "International Bans."

[489] United Nations Educational, Scientific and Cultural Organization, *"Great."*

[490] Messenger, "Zoos."

[491] Ivory, "Chimpanzee."

[492] O'Carroll, "Spain."

[493] Catalan, "Apes."

[494] Project R&R. "International."

[495] Ibid.

[496] Ynterian, "Zoonit."

[497] Govtrack.us, "H.R. 1326."

[498] Capital News Service, "Deficit."

[499] Project R&R, "Great."

[500] Gorman, "Agency."

[501] Thompson, *Intimate,* 232.

[502] Singer, *Animal,* xiii.

[503] International Primate Protection League, "History."

[504] Blum, *Monkey,* 120.

[505] International Primate Protection League, "History."

[506] Ibid.

[507] Ibid.

[508] Ibid.

[509] Ibid.

[510] International Primate Protection League, "History."

[511] Born Free Foundation, "Born."

[512] Born Free Foundation, "About."

[513] Born Free USA, "API's."

[514] Born Free USA, "Sanctuary."

[515] Born Free Foundation, "Born."

[516] Animal Legal Defense Fund, "About."

[517] Animal Legal Defense Fund, "Animal Legal."

[518] Student Animal Legal Defense Fund, "Law."

[519] Animal Legal Defense Fund, "Animal Law."

[520] Ibid.

[521] Animal Legal Defense Fund, "Landmarks."

[522] Ibid.

[523] Ibid.

[524] Conn, "Legal."

[525] Peterson, *Jane,* 98-100.

[526] The Jane Goodall Institute, "Early."

[527] Ibid.

[528] Ibid.

[529] Ibid..

[530] The Jane Goodall Institute, "Gombe."

[531] Quammen, "Jane."

[532] The Jane Goodall Institute, "Rescuing."

[533] The Jane Goodall Institute, "About."

[534] Borneo Orangutan Survival, "10."

[535] Orangutan Outreach, "About."

[536] Orangutan Outreach, "About."

[537] Miles, "Language," 43.

[538] ibid., 45.

[539] Grehan, "Mona."

[540] Orangutan Outreach, "Nyaru."

[541] Borneo Orangutan Survival, "Mawas."

[542] Borneo Orangutan Survival Australia, "Samboja."

[543] Orangutan Outreach, "Nyaru."

[544] Fossey, *Gorillas.*

[545] The Dian Fossey Gorilla Fund International, "Dian Fossey."

[546] The Dian Fossey Gorilla Fund International, "Karisoke."

[547] Ibid.

[548] The Dian Fossey Gorilla Fund International, "GRACE."

[549] The Dian Fossey Gorilla Fund International, "Fossey Fund's."

[550] The Dian Fossey Gorilla Fund International, "Karisoke."

[551] The British Union for the Abolition of Vivisection, "History."

[552] The British Union for the Abolition of Vivisection, "Achievements."

[553] The British Union for the Abolishment of Vivisection, "PACE."

[554] The British Union for the Abolishment of Vivisection, "PACE."

[555] New England Anti-Vivisection Society, "About."

[556] Project R&R, "End."

[557] Orangutan Foundation International, "Dr."

[558] Ibid.

[559] Orangutan Foundation International, "History."

[560] Brubaker, "Mission."

[561] Ibid.

[562] Orangutan Foundation International, "Rehabilitation."

[563] Thompson, *Intimate,* 232.

[564] Orangutan Foundation International, "Dr."

[565] Thompson, *Intimate,* 76.

[566] Thompson, *Intimate,* 222.

[567] American Sanctuary Association, "About."

[568] Ivory, "Chimpanzee."

[569] Ibid.

[570] Global Federation of Animal Sanctuaries, "Global."

[571] Ibid.

[572] Association of Zoos and Aquariums, "Health."

[573] Association of Zoos and Aquariums, "Becoming."

[574] Association of Zoos and Aquariums, "List."

[575] United States Department of Agriculture Animal and Plant Inspection Service, "Animal."

[576] Cavalieri and Singer, *Great,* 5.

[577] Science Daily. "Teeth."

[578] Science Daily. "Discovery."

[579] Stauffer, Walker, et al. "Human."

[580] Peterson and Goodall, *Visions,* 80.

[581] Ibid., 15.

[582] Ibid., 4.

[583] Ibid., 17.

[584] De Waal and Lanting, *Bonobo,* 8.

[585] Peterson and Goodall, *Visions,* 17.

[586] Ibid.

[587] Blum, *Monkey,* 41.

[588] Groves, "Why.".

[589] Corbey, "Ambiguous," 130.

[590] The Ram Dass Library, "Hanuman."

[591] Sumatran Orangutan Society, "Orangutan."

[592] De Waal and Lanting, *Bonobo*, 134.

[593] Thompson, *Intimate,* 43.

[594] Corbey, "Ambiguous," 129.

[595] Ibid., 132.

[596] Cavalieri and Singer, *Great,* 12.

[597] Rees, "Reflections."

[598] Ibid.

[599] Bekoff, "Common," 105.

[600] De Waal and Lanting, *Bonobo*, 34.

[601] Ibid., 33.

[602] Cavalieri and Singer, *Great,* 13.

[603] Thompson, *Intimate,* 84.

[604] Wise, *Rattling,* 126.

[605] Bekoff, 62.

[606] NOVA, "Ape," 52:45.

[607] Wise, *Rattling,* 165.

[608] Corbey, "Ambiguous," 134.

[609] Linzey and Morrison, "Is It?"

[610] Cavalieri and Singer, *Great,* 1.

[611] Singer, *Animal,* 6.

[612] Ibid., 11.

[613] Singer, *Animal.*

[614] Francione, *Animals, Property,* x.

[615] Wise, *Rattling,* 85.

[616] Cavalieri and Singer, *Great,* 5.

[617] Diamond, "Third," 95.

[618] Cavalieri and Singer, *Great,* 4.

[619] Singer, *Animal,* xiii.

[620] United States Department of Agriculture, "Environmental."

[621] Blum, *Monkey,* 189.

[622] Peterson and Goodall, *Visions,* 229.

[623] Save The Chimps, "Rescuing."

[624] Blum, *Monkey,* 210.

[625] Ibid., 6.

[626] Ibid., 46.

[627] Ibid., vii.

[628] Stanford, *Significant,* 198.

[629] Thompson, *Intimate,* 241.

[630] Green and The Center for Public Integrity, *Animal,* xxviii-xxix.

[631] Ibid., 62.

[632] Ibid., 14.

[633] Wise, *Rattling,* 13.

[634] Rollin, "Ascent," 217.

[635] Thompson, *Intimate,* 132.

[636] Dunbar, "What's," 110.

[637] Wise, *Rattling,* 139.

[638] Ibid.

[639] Mitchell, "Humans," 238.

[640] Miller, "Wahokies," 230-237.

[641] Clark, "Apes," 124.

[642] Rachels, "Why," 155.

[643] Häyry, and Häyry. "Who's," 174.

[644] Ibid., 178.

[645] Ibid., 181.

[646] Cavalieri and Singer, *Great,* 6.

[647] Rollin, "Ascent," 209.

[648] Francione, *Animals as Persons,* x.

[649] Francione, *Animals, Property,* 101.

[650] Beers, *Prevention,* 3.

[651] Wilcox, *Voice.*

[652] Rachels, "Why," 153.

[653] Ibid., 156.

[654] Jefferson, Thomas, Letter.

[655] Tanner, *Voices.*

[656] Fouts and Fouts. "Chimpanzees'," 29.